A Decolonial Black Feminist Theory of Reading and Shade

This book uses a decolonial Black feminist lens to understand the contemporary significance of the practices and politics of indifference in United States higher education. It illustrates how higher education institutions are complicit in maintaining dominant social norms that perpetuate difference. It weaves together Black feminisms, affect and queer theory to demonstrate that the ways in which human bodies are classified and normalized in societal and scientific terms contribute to how the minoritized and marginalized feel White higher education spaces. The text espouses a Black Feminist Shad(e)y Theoretics to read the university, by considering the historical positioning of the modern university as sites in which the modern body is made and remade through empirically reliable truth claims and how contemporary knowledges and academic disciplinary inheritances bear the fingerprints of racist sexist science even as the academy tries to disavow its inheritance through so-called inclusive practices and policies today.

This book will appeal to students and scholars interested in Black feminism, Gender and women's studies, Black and ethnic studies, sociology, decoloniality, queer studies and affect theory.

Andrea N. Baldwin is an Assistant Professor of Women's and Gender Studies and Africana Studies in the Department of Sociology at Virginia Tech, USA.

Routledge Research on Decoloniality and New Postcolonialisms
Series Editor: Mark Jackson
Senior Lecturer in Postcolonial Geographies, School of Geographical Sciences,
University of Bristol, UK

Routledge Research on Decoloniality and New Postcolonialisms is a forum for original, critical research into the histories, legacies, and life-worlds of modern colonialism, postcolonialism, and contemporary coloniality. It analyses efforts to decolonise dominant and damaging forms of thinking and practice, and identifies, from around the world, diverse perspectives that encourage living and flourishing differently. Once the purview of a postcolonial studies informed by the cultural turn's important focus on identity, language, text and representation, today's resurgent critiques of coloniality are also increasingly informed, across the humanities and social sciences, by a host of new influences and continuing insights for different futures: indigeneity, critical race theory, relational ecologies, critical semiotics, posthumanisms, ontology, affect, feminist standpoints, creative methodologies, post-development, critical pedagogies, intercultural activisms, place-based knowledges, and much else. The series welcomes a range of contributions from socially engaged intellectuals, theoretical scholars, empirical analysts, and critical practitioners whose work attends, and commits, to newly rigorous analyses of alternative proposals for understanding life and living well on our increasingly damaged earth.

This series is aimed at upper-level undergraduates, research students and academics, appealing to scholars from a range of academic fields including human geography, sociology, politics and broader interdisciplinary fields of social sciences, arts and humanities.

A Decolonial Black Feminist Theory of Reading and Shade
Feeling the University
Andrea N. Baldwin

Transdisciplinary Thinking from the Global South
Whose problems, whose solutions?
Edited by Juan Carlos Finck Carrales and Julia Suárez-Krabbe

For more information about this series, please visit: https://www.routledge.com/Routledge-Research-on-Decoloniality-and-New-Postcolonialisms/book-series/RRNP

A Decolonial Black Feminist Theory of Reading and Shade

Feeling the University

Andrea N. Baldwin

Routledge
Taylor & Francis Group

LONDON AND NEW YORK

First published 2022
by Routledge
2 Park Square, Milton Park, Abingdon, Oxon OX14 4RN

and by Routledge
605 Third Avenue, New York, NY 10158

*Routledge is an imprint of the Taylor & Francis Group,
an informa business*

© 2022 Andrea N. Baldwin

British Library Cataloguing-in-Publication Data
A catalogue record for this book is available from the British Library

Library of Congress Cataloging-in-Publication Data
Names: Baldwin, Andrea N., author.
Title: A decolonial Black feminist theory of reading and shade: feeling
    the university/Andrea N. Baldwin.
Description: Abingdon, Oxon; New York, NY: Routledge Books, 2022. |
Series: Routledge research on decoloniality and new postcolonialisms |
    Includes bibliographical references and index.
Identifiers: LCCN 2021021390 (print) | LCCN 2021021391 (ebook) |
    ISBN 9780367894801 (hardback) | ISBN 9781032118765 (paperback) |
    ISBN 9781003019442 (ebook)
Subjects: LCSH: Racism in higher education–United States. | Sexism in
    higher education–United States. | Feminism and higher education–
    United States. | Education, Higher–Social aspects–United States.
Classification: LCC LC212.42 .B35 2022 (print) | LCC LC212.42 (ebook) |
    DDC 379.2/6–dc23
LC record available at https://lccn.loc.gov/2021021390
LC ebook record available at https://lccn.loc.gov/2021021391

ISBN: 978-0-367-89480-1 (hbk)
ISBN: 978-1-032-11876-5 (pbk)
ISBN: 978-1-003-01944-2 (ebk)

DOI: 10.4324/9781003019442

Typeset in Times New Roman
by KnowledgeWorks Global Ltd.

Contents

Acknowledgments

This book would not have been possible without the love, dedication, support and assistance of a number of people.

Amy Splitt, whose editing and suggestions forced me to keep mining, to stick to my reading practice and to be shady as fuck! This text is a result of Amy's generosity and friendship.

My amazing friend, colleague, and writing partner Nana Brantuo who kept me sane throughout this process as she checked on me every day, listened to me talk about the project, about life, about the feeling of futility writing a book in the midst of the pandemic, who was with me at my most excited and also at my lowest points in writing this book. She was the constant that I needed throughout this process, always reminding me, "you got this sis!" providing the right amounts of love and support I needed.

My undergraduate research assistant Rosie Banks who was also crucial to the writing and editing process. Rosie is such an amazing and talented researcher. Her dedication to this project was in helping me work through my citations and references.

The research participants for both the Posture Portraits and the Self Care study and whose narratives were integral to this book and appear in abbreviated form in Chapter 3. Without them, this project would not be possible.

Other colleagues who helped me through this process by either reading a chapter, listening to a talk, helping me with a presentation, searching for a cultural artifact, generally encouraging me, and just always being there with suggestions and care. They include my colleagues from the Coast writing group at Virginia Tech, Kimberly Williams, Cherise Harris, Jennifer Nival, Trichia Cadette, Paula Seniors, Shelley Bayne and many, many others.

My sister Trenicia Griffith for always being ready with a distraction and for giving me the joy and inspiration that is my niece Ava-Grace.

Finally, Quentin and Liam Baldwin, my husband and son, and my mother Sandra Griffith. These three are the relations from which my poetics spring forth.

Introduction

At the fall 2019 orientation at Virginia Polytechnic Institute and State University (Virginia Tech) the order of business was, as it has become customary at universities and colleges across the United States in the last decade or so, that all participants including incoming students, would indicate their names and pronouns on their orientation badges. The orientation speaker began their speech by acknowledging, "we are standing on stolen lands," as a means of demonstrating that the institution understood the importance of this fact as part of its performance of inclusion and diversity. Like most higher education institutions (HEIs) in the United States today, particularly larger predominantly white institutions (PWIs), Virginia Tech has instituted and invested in offices of diversity and inclusion, hired people (usually Black and Brown) to staff these offices, and designed and implemented policies and procedures to promote diversity, including to work on making orientation a more welcoming space for those with minoritized and marginalized identities.

Following the 2019 orientation to welcome the newest Hokies, the university found itself on the receiving end of some very public backlash. On August 14th, a few days after the orientation, a disgruntled parent of an incoming first-year Reserve Officers' Training Corps (ROTC) student, published an article on the conservative online platform *the Federalist*. In the article entitled "My Son's Freshman Orientation At Virginia Tech Was Full Of Leftist Propaganda"(Nance 2019), the parent, Penny Nance, inquires "Are taxpayers funding academic institutions to indoctrinate our kids?" In the article, she describes what she terms "madness," that is "overtly leftist propaganda spewed at ... orientation" that not only didn't "remember the names of fallen cadets on the pylons or the 32 dead and 17 injured in the 2007 Virginia Tech campus shooting," but rather made "the stunning choice to open orientation by recognizing two Native American tribes on whose land the college was built (with the implication that it was stolen)." Nance goes on to state (and illustrates by a photo of a name badge, presumably her son's showing the pronouns "He, Him, His, Himself"), that for the next two hours of the orientation, "speaker after speaker ... introduced themselves with not just their names and titles but also preferred pronouns" and described one

DOI: 10.4324/9781003019442-1

of the presentations as a "heavy-handed diversity lecture." In the remainder of the article, Nance describes other parts of the orientation experience with statements like, "they lectured students on not making assumptions about each other's gender or sexuality. ... showcased identity group politics, but certainly not all identities. It's apparently way cooler to be a minority trans woman with food allergies than simply to be an American college student." She concluded her article with a call to action to all Virginians to get up and do something. She writes,

> As a mom, part of me wanted to load my son in the car and head up the road to Liberty University ... Why should conservative kids be forced to become educational refugees from public institutions that, despite enjoying our tax dollars, don't welcome us? ... Virginia Tech and most other public universities have forgotten they work for us. ... Virginians deserve better. We do not bow to the ascendancy of the liberal, ivory-towered academic's worldview over ours (Nance 2019).

It is for several reasons that I begin this book, about Black feminism, decoloniality, and higher education in the United States (US), with the above example. First, this book is about examining how the colonial roots of U.S. higher education have meant that certain groups of people have traditionally felt entitled to occupy the spaces and utilize the resources of the academy. The U.S. college student, the "us" that Nance invokes, is read as white, middle-class, Christian, cisgender, heterosexual, and mostly male from a demographic group who believes that attending university is their birthright. As we know, the United States has historically produced forms of governmentality which according to Wendy Brown – who is writing about nation states generally – "produc[e] who the 'we' is: who's in, who's out, who's needed, who's not needed, identities that are racialized, ethnicized, and "religionized," sometimes in incoherent yet consequential ways. ... There are forms of policing, securitizing, categorizing and identity-making that saturate the internal lives of nations engaged in them, and that do not just happen at their borders" (Celikates and Jansen 2013, n.p.). Recognizing this internal formation of nation states is key, as the "us" Nance describes as "American" also defines the "not us," those who have traditionally been excluded from the university and whose ancestors' labor and lands were integral to its building.

Nance's example also demonstrates the import of *affectivity* produced as a result of the type of relation one has to the university. Nance communicated in her article a feeling of nonbelonging, and the wanting, longing and desire to be a part of and included, recognized, and to have her feelings validated. In many of the writings on U.S. higher education, the engagement with affect is central to what may be described as "the college experience" – which contributes to the continued value of higher education to generations of families and students, even considering what appears to be a low return of their financial investments, as student tuition and student loan debt have skyrocketed. However, this central "why" in the discussion of how students in this present COVID-19 pandemic are willing to risk their lives to return to campuses across the nation, is largely absent. When writing about bodies, particularly around issues of injustice, it is important that we engage with the affective relationships that people develop around not only institutions but also around ideas about what gaining access to, belonging to, being a part of, and included in, these institutions means, and conversely what it means to be left out, unacknowledged, to feel as though one does not belong, that one is a stranger, and to be anxious about what that means for one's life and future.

Sarah Ahmed writes that "some-bodies are more recognizable as strangers than other-bodies precisely because they are already read and valued in the demarcation of social spaces"(2000, 20). Nirmal Puwar also makes this point in her 2004 text *Space Invaders: Race, Gender and Bodies Out of Place*. She writes

> Formally, today, women and racialised minorities can enter positions that they were previously excluded from … However, social spaces are not blank and open for any body to occupy. There is a connection between bodies and space, which is built, repeated and contested over time. While all can, in theory, enter, it is certain types of bodies that are tacitly designed as being the 'natural' occupants of specific positions. Some bodies are deemed as having the right to belong, while others are marked out as trespassers, who are, in accordance with how both spaces and bodies are imagined (politically, historically and conceptually), circumscribed as being 'out of place'. Not being the somatic norm, they are space invaders. The coupling of particular spaces with specific types of bodies is no doubt subject to change; this usually, however, is not without consequence as it often breaks with how bodies have been placed (8).

The colonial history of the academy that assured Nance that she and her son belong to the academic social space at Virginia Tech, that she could claim and own it, navigate it "moving within, or passing through" (Ahmed 2000, 32), juxtaposed with her feelings of nonbelonging and being made to feel unwelcome is what spurred Nance's call to action to Virginians to rise up against the "madness" she witnessed. That Nance refers to what she experienced as madness, while ableist and offensive, is no coincidence. Black feminist Katherine McKittrick in her work on Black geographies has written about "rational spatial colonization and domination" which profit from the concealment and "erasure and objectification of subaltern subjectivities, stories, and lands" (McKittrick 2006, x). McKittrick writes about the ways in which the process of spatialization have "seemingly predetermined stabilities, such as boundaries, color-lines, "proper" places, fixed and settled infrastructures and streets, oceanic containers" (2006, xi). She argues, however, that these spaces/geographies are not fixed and secure but rather that they produce and are produced socially through meaning. Space isn't just is, rather it is premised on "[c]oncealment, marginalization, boundaries …[as] important social processes. We make concealment happen; it is not natural but rather names and organizes where racial-sexual differentiation occurs" (2006, xi–xii). Here I revisit the question of "We."

Nance's personally and politically disorienting encounter at "orientation" in a space that has historically worked diligently to conceal and objectify certain bodies through rational spatial colonization and domination, but which is now publicly and performatively recognizing those bodies as part of "us and we" was both maddening and madness to Nance as the boundaries she had taken for granted appeared unfixed and wavering. Nance felt out of place as she was per Ahmed "with others who are already [and should be] recognized as stranger(s), as out of place in this place" and this feeling produced "forms of discomfort and resistance, … felt on the skin" (2000, 50). This discomfort drove Nance to write that Virginia Tech, "despite enjoying our tax dollars, [does not]… welcome us." She seeks validation in her familial appeal to other (white conservative) Virginians who may also feel that what is theirs, what they have and continue to pay for and support, is slowly being taken from them, so much so that they now find themselves on the periphery and begin to panic, wishing to load their children up in the car and head over to Liberty University. This is a moral panic which Stuart Hall points out is an ideological process. It

> represents a way of dealing with what are diffuse and often unorganised social fears and anxieties. It deals with those fears and anxieties, not by addressing the real problems and conditions which underlie them, but by projecting and displacing them on to the identified social group. That is to say, the moral panic crystallises popular fears and anxieties which have a real basis, and, by providing them with a simple, concrete, identifiable, simple, social object, seeks to resolve them. Around these stigmatised groups or events, a powerful and popular groundswell of opinion can often be mustered. It is a groundswell which calls, in popular accents, on the 'authorities' to take controlling action (2017, 154).

Nance's anxiety of disorientation led to an appeal to the "us" as she projected her anxieties on the "not us." Nance engages in what Hall refers to as "the language of moral panics," as a way of explaining through popular and political terms her discontent to the power (2017, 154) bloc of white conservative Virginians. White Americans like Nance know far too well that the language of moral panics has the power to realign spaces, bringing them back from the brink of so-called "madness" to a more rational spatialization.

But Nance's feelings must also be juxtaposed with those of the historically unwelcomed whom the university is now attempting to make *feel* welcomed. Those feelings/disruption of affectivity provoked via welcoming (mostly through verbalized) performativity demonstrate how these relatively novel university diversity and inclusion initiatives upset those traditionally included and who now feel like these diversity and inclusion measures are working against them. Examining this pushback leads to a closer analysis of the performativity of inclusion and diversity at the university.

Considerable established literature critiques the university's performance of diversity as largely about perception (Ahmed 2012). Like Sarah Ahmed, mentioned above, Iris Marion Young, in her 1990 text, *Justice and the Politics of Difference,* refers to the commitment to formal equality as tending "to support a public etiquette" which "demands that we relate to people as individuals only, according everyone the same respect and courtesies" (132). Performing diversity and inclusion as a means to manage perception or in the vein of exhibiting good manners has serious implications. Without institutional change, superficial acknowledgment can put those with minoritized and marginalized identities in grave danger. This hazard comes not only in the form of explicit revolt like Nance's. More Subtle threats continue despite "diversity and inclusion." One can be well-mannered and *treat people nice* while also providing little help or support, refusing to acknowledge issues that negatively impact marginalized persons; and in fact, as the saying "Bless your heart" in the southern U.S. parlance demonstrates, actively wishing them ill with a smile. In the university, these courtesies send a strong signal to those on the receiving end that they do not belong.

Nance's call to action in *the Federalist* publicly announced that the university belonged to "us" – and by "us" she is referring to conservative Virginians who "do not bow to the ascendancy of the liberal, ivory-towered academic's worldview over ours." While the university is not responsible for Nance's opinions, the university is responsible for the way in which it decides to engage or not with such public critique and propaganda, and how it works to make its students, faculty and staff feel supported. In this instance, the university's response was predictably anemic. Those confident in their belonging, mostly white and male, generally dismissed Nance's article as drivel with flippant statements like "well I guess this means the semester has officially begun"; while those with minoritized and marginalized identities – the "not us" – perceived the threat in Nance's call to action.

As colleagues on my own department's listserv were sharing their rather dismissive opinions of Nance, I reached out privately to the then-chair of my department to register the effect that the article had on me: a Black woman, immigrant, junior faculty who teaches classes which can be considered as Nance describes "left leaning" – queer studies, Black studies, and Black feminisms – and who clearly is not part of the "us" Nance was advocating for in her article. I requested a department-level intervention where we could have candid conversation about Nance's article and the university's silence. To add context, the previous semester, in my Introduction to African American studies class a white male student tried to intimidate both me and the teaching assistant and was offensive to other students in a majority Black classroom. While this student eventually withdrew from the class, I was rattled. From that standpoint, I was afraid that an article like Nance's would embolden those who saw the classes I taught as propaganda. I legitimately feared Black students and I would be targeted yet again. Not only was I terrified by the possibility of violence; I was appalled at the flippant way that colleagues were engaging the matter of Nance's article, and the nonresponse from the university. There was no engagement with how the article had potential real negative consequences – emotional, psychological, physical and professional – for those on campus who could be targeted for harassment and worse in the cis-heterosexist culture, and the current anti-Black, anti-immigrant climate.

Responding to my request, the chair stated that while he recognized the article as symptomatic of a larger trend of efforts by conservative groups to delegitimize universities, there was little the university could do to counter this broader trend. He assured me that the department, college, and university all supported the principles of academic freedom, and that the university had a long history of supporting professors accused of "spreading propaganda" or similar. He then suggested that I personally bear the burden of devising a strategy to avoid incidents in my class. For example, I could include a statement in my syllabus about the university being a place for the free exchange of ideas in an atmosphere of civility, with language copied verbatim from the Student Handbook on "student rights and responsibilities" and the importance of following the "Principles of Community." I was told that I was within my rights to initiate disciplinary action ranging from simply dismissing a disruptive student from class and voiding his or her attendance for that day, to reporting the student to the Dean of Students for disciplinary action including dismissal from the university, though he acknowledged that any punishment of that sort would be unlikely unless the case was considered very serious or the student had a prior record. The correspondence also stated that if at any time I believed a student to be threatening and potentially dangerous, I should inform the university's threat assessment team. The then-chair also suggested holding a workshop for faculty on how to deal with potentially disruptive or dangerous students in the class and assured me that he would relay my concerns to the Office of Inclusion and Diversity. I felt that my own safety and that of marginalized students was left up to me, with only policy, not power, to back us up.

This response, to use a popular phrase, "had me in my feelings." I was furious, disappointed, and sad all at the same time. "When diversity becomes a form of hospitality" then it produces certain feelings, as those welcomed are incorporated as guests. Guests by definition are contingently present, are supposed to be gracious (and grateful). As such, my actions in speaking out about my feelings, to call out whiteness, was read "as a sign of ingratitude, of failing to be grateful for the hospitality ... received by virtue of ...arrival" (Ahmed 2012, 43).

The language of tolerance and civility offered the department and the university a means of avoiding the responsibility of doing the actual work of making the campus a welcoming and safe place. Furthermore, it erased the lack of power that I and certain other bodies automatically show up with. The identities which make us targets prevent us from having the perceived authority to do what the guidelines state (especially understanding how they will be perceived by students and how this perception could have negative impacts on career trajectories). Clearly, the suggested workshop, while it would "make white subjects feel good" (Ahmed 2012, 170) about their work to address discrimination, would not be sufficient to address the concerns that so many of us on campus have about how the ways we show up as strangers on campus endanger us within and outside of the confines of the classroom. I walk around campus in an immigrant body with a noticeable immigrant accent in an anti-immigrant climate. I also walk around in a Black body in a racist America while also teaching courses on Blackness. In addition to these, I also live in a woman's body in a place where the rhetoric of raping women of color – including an elected congresswoman (Thompson 2019) – is at best shrugged off and at worst condoned. Furthermore, I teach courses that promote the humanity of women and queer folx – the "propaganda" which white and conservative students publicly deplore. The university always has me in my feelings! And there are others who feel the same.

Universities are quick to bring our bodies to campus as a "sign of inclusion [that] makes the signs of exclusion disappear" (Ahmed 2012, 65), but they are slow to address issues which threaten and Other us. We are viscerally and personally *affected*, that is, *feel* the lack of institutional protection to back up this "inclusion." Consider: at public institutions, contact and other identifying information including our salaries is available as a matter of public record, in a climate where the impacts of former President Trump's rhetoric about immigrants stealing American jobs still remain. With just a few clicks, a disgruntled person can find out what we teach, what we make, and the location of our offices. Those who are "not us," who are seen as teaching and doing the very things Nance has claimed makes tax paying, deserving Virginians, American citizens (read: white people) feel unwelcomed, are *affected*. To see the diversity etiquette play out before your very eyes in the form of no university-wide statement or implementation of some concrete measures in the name of protecting our free speech, and then to have to interact with well-mannered colleagues who don't have to worry about their safety because they are part of "us," produces a particular affective relationship between the minoritized and the marginalized and these institutions, where we too feel, despite our inclusion, unwelcomed.

Whom does rhetoric of diversity and inclusion benefit? Those who do not suffer from racism get to feel good about telling those who do what should be done. They get to feel good and happy about their performances and solutions. But "then who gets to feel bad about racism" (Ahmed 2012, 170)? According to Ahmed: "happy whiteness, even when this happiness is *about* antiracism, is what allows racism to remain the burden of racialized others. Indeed ... bad feelings of racism (hatred, fear, and so on) are projected onto the bodies of unhappy racist whites [the Nances of this world], which allows progressive whites to be happy with themselves in the face of continued racism toward racialized others" (2012, 170).

The performance of diversity and inclusion without real support by these institutions makes space for the ensuing backlash, and places those who the university seeks to include in a position of potential harm. Take, for example, the university's anemic responses to the recent 2020 attempt by the former Trump administration to deport international students who because of their educational institution's policy response to COVID-19 may have had to take classes online; or the Trump administration's public charge policies which required international students "to demonstrate they have not made extensive use of public benefits if they wish to extend or change their visa status after they arrive" (American Council on Higher education 2019, n.p.) and thus prevented students on F-1 and J-1 visas at state-funded universities like Virginia Tech from accessing relief funds. Students who were unable to return to their homes abroad because of travel restrictions due to COVID-19 but also due to the Trump administration's ban on returning to the United States from some Muslim countries were in immediate need of financial aid to help them get through the hardships brought on by the pandemic. However, because they were suddenly ineligible for federal monies allocated to Virginia Tech (some $19 million) due to the public charge policies mentioned above, they could not access these funds; meanwhile, there were no clear policies on what funds they could access or how the institution would assist. International students "contributed $45 billion to the U.S. economy in 2018, according to the U.S. Department of Commerce" (iie 2019, n.p.). While the neo-liberal capitalist university welcomes this influx of revenue, in this time of global pandemic, these students were made to feel uncertain and exasperated as they struggled with multiple threats: the possibility of homelessness due to lack of resources, the potential for violent and traumatic situations due to COVID-19 as seen with the uptick in anti-Asian violence, and stirred up white U.S. nationalisms which have resulted in highly visible displays of racism, hate crimes, and anti-immigrant and anti-Black sentiment (Hundle 2019, 296). The ways in which international students then came to *feel* the university during this time, in other words, the *affectivity* of their situation became even more heightened as policies of the Trump administration demonstrated how precariously "people become habituated within institutions – how they come to occupy spaces that have already been given to them" (Ahmed 2012, 123)– as hosts or stranger.

In this contemporary moment, when universities are holding themselves out as bastions of diversity and inclusion, they still support and are supported by a system of colonial rationalization that marks some bodies as always out of place. As imbricated in a system of colonial domination, the university has produced and engaged in a set of colonial logics which seek the continuation of its survival as society changes over time. This survival now includes the calculated performance of diversity and inclusion as a matter of perception and etiquette to encourage groups formerly excluded to *feel* about and for the university in certain positive and affirming ways. However, to animate affective relationships and attachments that emerge from a yearning to attend and achieve the possibilities the university holds out, being truly considered part of the institutional "we" and "us" can be had only by being a part of it: true belonging, not contingent "hospitality," must exist beyond the performative language of policy as the crucial step in achieving the better life and future for which one yearns.

In this book, I examine the ways in which colonial logics of domination at U.S. universities have historically deployed a process of academic justification, including through data collection, to frame, reframe and articulate how these logics benefit those they seek to dominate and their futures. I demonstrate how these logics, in fact, work to collect data to advance a neoliberal capitalist order in which the university is a part. I am interested in how according to Sarah E. Chinn, we theorize bodies as evidence, how "concepts of evidence formed an explanatory network of systems that claimed to help one recognize different kinds of people, particularly in terms of race, through looking at their bodies" (2000, xv). I am interested in the historical performance of etiquette as inclusion as a means of managing perception through collecting "perception data"(Ahmed 2012) and information on and about marginalized and minoritized bodies in order to secure the future of the university as a transparent space with the numbers/evidence to prove its dedication to diversity even as it "make[s] concealment happen" (McKittrick 2006, xi-xii) by way of and as a result of this data.

Throughout this text, I examine how both racist and sexist exclusion "have a source in the structure of modern reason and its self-made opposition to desire, body, and affectivity" (Young 1990, 124) and how nineteenth and twentieth centuries "science established unifying, controlling reason in opposition to and mastery over the body, and then identified some groups with reason and others with the body" so that these bodies were excluded from the academy (Young 1990, 124). I also examine how the foundations of this science exist today in the ways in which inclusion of historically excluded bodies in academic spaces is more a reflection of, as Angela Davis observes, education as "a commodity ... so thoroughly commoditized that many people don't even know how to understand the very process of acquiring knowledge because it is subordinated to the future capacity to make money" (2016, 120). I extrapolate this commodification to the "inclusion" of Black/Brown/female/immigrant/queer/disabled bodies without a commitment to equity. The performance of diversity and inclusion, in fact, has nothing to do with these bodies but everything to do with capital and the university as a capitalist institution, starting from its colonial foundations.

These bodies themselves are a commodity. I offer the analogy of manufacturing: fed through an assembly line, the material that enters is transformed into a particular product for some specific use. Innovation may be pursued for various reasons: because of a need for efficiency or a change in societal habits, perceptions, and tastes and ethics. Producers need a way of collecting and analyzing data so that they can better understand these changes and cater to them, thus remaining competitive and profitable. I observe how the university is using the performance of diversity and inclusion strategically and dispassionately as an innovation tool in its assembly line as it seeks to stay economically competitive in a society that is rapidly becoming less white.

If we think of diversity and inclusion as a tool to measure and quantify bodies in the service of capital, then we can understand why even as the university is performing its diversity work, those whom they admit still feel in many ways as though they do not belong. I argue that the issues those with minoritized and marginalized identities who enter the university continue to face due to the university's performance of diversity and inclusion stems from what I am referring to as a *politics of indifference,* that is, that the inclusion of minoritized and marginalized bodies in no way stems from an actual real interest in them as members of the community. Rather diversity itself is tied to the university's colonial, racist, sexist legacy, which historically has used and exploited these bodies for the benefit and furtherance of capitalist society, while also being actively indifferent to the ways in which the institution uses these bodies and the consequences that stem therefrom and that negatively impacts said bodies.

In *Scenes of Subjection: Terrorism Slavery, and Self-making in Nineteenth-Century America*, Saidiya Hartman writes about what she calls "the precariousness of empathy and the uncertain line between witness and spectator" (1997, 4). She writes about the obscene display of slavery's brutality and the tortured bodies of the enslaved and asked, "how does one give expression to these outrages without exacerbating the *indifference of suffering* that is the consequence of the benumbing spectacle or contend with the narcissistic identification that obliterates the other or the prurience that too often is the response to such displays" [emphasis mine] (4)? In this text, it is this historical indifference of or to the suffering caused by practices of racism and sexism, which I examine. Like Hartman, I am wary of exacerbating the indifference of suffering by focusing voyeuristically on the bodies of the marginalized and minoritized but rather seek to examine the indifference of/to suffering as a political act. In other words, what I am calling the politics of indifference – the seemingly detached, apathetic, well-reasoned and systematic approaches and responses used by universities surrounding considerations of gender, race/ethnicity (sexuality, ability, class), inclusion and exclusion, and how these policies lead to university-sanctioned practices which have traumatic consequences for, in the case made in this text, women and racialized bodies – is the way in which the university presumes/pretends/portends that these bodies are and should *feel* content with hospitality and inclusion etiquette. Riffing off of Hartman again, it is "the coerced enactment of indifference and the orchestration of diversions" (1997, 38). By connecting current institutional policies of inclusion in this way to the ways in which the university has historically used said bodies to produce itself - their land, labor, and bodies as data - it is clear how inclusion and exclusion are two sides of the same coin. It is clear that the same indifference, described by Hartman, with which these bodies were made to dance and sing as they were exploited to supply the raw material for the benefit of U.S. capitalism and the academy, is the same indifference experienced by those included today in the academy as part of advance capitalism in its contemporary form.

I came to think about the ways in which universities perform diversity and inclusion as stemming from this place of indifference while doing research on two separate projects on gender and race in higher education. I became intrigued at how even despite claims of diversity and inclusion, the academy remained quite wedded to its racist/sexist/heteronormative genealogy. I became interested in how, according to Sarah Ahmed, the act of inclusion could be read as an act to maintain exclusion (2012, 43). These two nonrelated projects; one on the posture portraits, a practice of taking nude and seminude photographs of incoming first-year students across the United States during the mid to late 1800s to the 1960s, the other examining self-care for Black and Brown students at PWIs, a project which I started in 2016, made me think more about Ahmed's statement. I thought about how while today universities are working to include more Black and Brown faculty, staff, and students, women, and queer folx, to make "the signs of exclusion disappear" (Ahmed 2012, 65), that these signs in many cases – as shared with interview participants for the self-care study – were becoming more pronounced and the negative effects of this inclusion and diversity work was being felt profoundly, and precluded students from caring for themselves. These two studies made me curious about how in the current time of diversity and inclusion in the academy students, Black and Brown students in particular, were echoing the same sentiments as the white upper-middle-class women who I interviewed about their experiences of inclusion/exclusion through the posture portraits in the 1940s, 1950s and 1960s. How were two university practices, the posture portraits and diversity and inclusion, both propagated by the university decades apart, focused on inclusion/exclusion, evoking similar affective responses from participants – feelings of frustration, embarrassment, and of nonbelonging?

The politics of indifference, I argue, is deployed through what I refer to as an *apathetic methodology of power*, that is the commitment by those in decision making positions at colleges and universities to value policy, science, and so-called academic rigor, unconcerned with the real-life experiences of those whom mere policy does not in fact protect. The *apathetic methodology of power* harms the affective relationship between the minoritized and marginalized and the university, while at the same time binding so-called inclusion to accepting the dispassionate politics of institutional power, which functions only to sustain itself. Returning to Hartman, who writes about the captive body yoked "to the will, whims, and exploits of the owner and by the constancy of the slave's unmet yearnings, whether for food or for freedom" (1997, 49), it is important to recognize how under the worst possible circumstances desire can produce what Hartman refers to as "consenting agency," that is "a range of everyday acts, seemingly self-directed but actually induced by the owner" (1997, 54) and as I argue, those with power in the university. To further explore the ways in which the minoritized and marginalized feel the university, I borrow from bell hooks' work on yearning (1990) to work through what I call *the promise and politics of yearning* - which I will come back to later. I ask, how have bodies been constructed by so-called objective modern academic (political and scientific) practices, how has this construction determined whether, when, and under what circumstances one is included, and whether this determination has implications for how bodies feel the university?

I use the personal narratives of interview participants from the two studies mentioned above to develop a socio-historical decolonial Black feminist analysis, signifying how the continued categorization of bodies by the university as normal or different, demonstrates how the politics of indifference has and continues to work to exclude even when one is included, and to make the case that we should be critical of inclusion arguments based on and *in* difference. The narratives highlight how institutions collect data for a particular purpose unrelated to the wellbeing of the people whose data they are collecting using them as instruments for the larger project of showing/ proving that the university was being managed and was managing its constituents/commodities in the best interest of society at large. This data could be used to demonstrate the value of the university as a place at the front-lines of society, an arbiter and authority of knowledge through an experimentation on Others. Especially in times of skyrocketing tuition and pervasive societal changes, the university needs to continue to solidify its position as the place of knowledge even as it continues to be challenged by those who through its knowledge production it has excluded – women, folks of color, queer folx, and others. It must absorb these bodies as and through data and use that data to make/turn bodies into things (re)counted/(re)measured for inclusion or exclusion based on this (re)counting/(re)measuring.

In this text, I explore the above claims using a decolonial Black feminist lens to engage what Rinaldo Walcott calls "diaspora reading practices" (2003) of the genealogy of the university, with the purpose of telling a story which counters the current diversity and inclusion rhetoric being peddled by the university. I demonstrate how the yearnings of the minoritized and marginalized for a better life and time is valuable in several ways, particularly in "its facilitation of collective identification ... And yet ... [it is] ensnared in a web of domination, accumulation, abjection resignation, and possibility" (Hartman 1997, 49) linked to white feelings of goodness, pleasure, benevolence and indifference. Hartman writes of pleasure, "It was nothing if not cunning, mercurial, treacherous, and indifferently complicit with quite divergent desires and aspirations, ranging from the instrumental aims of slave-owner designs for mastery to the promise and possibility of releasing or redressing the pained constraints of the captive body. It is the ambivalence of pleasure and its complicity with dominative strategies of subjection ..." (1997, 49–50) that perpetuates oppression. A diaspora reading practice then, involves being able to be as astute as the cunning, and to account for what is static in the mercurial. These types of reading practices according to Walcott "are the histories, memories, desires, free associations, disappointments, pleasures and investments we bring to any given texts ... informed by the peregrinations of ... consciousness" (2003, 118).

In this text, I use the examples mentioned above to also *read* the textuality of the university in the same way that Dorian Cory in the 1991 documentary *Paris Is Burning* (Livingston) defines a queer reading. In the film, Corey states, "shade comes from reading. Reading came first. You get in a smart crack, and everyone laughs and kikis because you've found a flaw and exaggerated it, then you've got a good read going. Shade is I don't tell you you're ugly but I don't have to tell you because you know you're ugly ... and that's shade." Cory's statement that a good read is to find a flaw and exaggerate it is the type of reading engaged in this text. Herein, I expose the flaws, meaning the cracks in the fictional image generated by the academy through the use of perception data they collect, by using a decolonial Black feminist reading practice or what I am referring to as a *Black feminist shad(e)y theoretics*. This read stems from and privileges the affective in "an attempt to exercise agency, as a willful form of territorial exertion in the service of autonomy, but one that is frustrating and frustrated" (Nash 2019, 28). This reading is personal and as such is reflexive and goes beyond a mere retelling of a history of the U.S. nation state project from which the academy originates and is complicit. Rather, this reading is informed according to Walcott by desires – what I have referred to herein as yearnings – and disappointments, is tentative and attempts to situate "struggle within ... history and [hence is able] to assert other and different kinds of responses" and that is transgressive (2003, 118).

By a decolonial Black feminist lens, I mean a lens that seeks to critically examine "the racial, gender, and sexual hierarchies that were put in place or strengthened by European modernity as it colonized and enslaved populations throughout the planet" (Maldonado-Torres 2007, 2–3). I draw from the work of several decolonial Black feminist scholars, including Carol Boyce-Davies and Katherine McKittrick. The definition of Black feminism deployed here is taken from Jennifer C. Nash's 2019 text *Black feminism reimagined* worth citing here at length. Nash writes,

> I treat black feminism as a varied project with theoretical, political, activist intellectual, erotic, ethical, and creative dimensions; black feminisms is multiple, myriad, shifting, and unfolding. To speak of it in

the singular is always to reduce its complexity, to neglect its internal debates and its rich and varied approaches to questions of black women's personhood. I treat the word "black" in front of "feminism" not as a marker of identity but as a political category, and I understand a "black feminist" approach to be one that centers analyses of racialized sexisms and homophobia, and foregrounds black women as intellectual producers, as creative agents, as political subjects, and as "freedom dreamers" even as the content and counters of those dreams vary (5).

This definition is decolonial because it permits us to engage the varied ways in which hegemonic power impacts us, to imagine ourselves as agents outside of the repressive structures created by this power, and to "envision and demand ... space where ... agency, autonomy, and self-determination would be at the center" (Alexander and Mohanty 1996, xli). This lens challenges the dominance of particular knowledges and knowledge practices, rhetoric, and power. A decolonial Black feminist lens "differs from postcolonialism because it unsettles the concept of colonialism in the first place; whereas postcolonialism is the "after" of colonization, decoloniality is the liberation from colonial structures, including values, methods, and knowledges" (McLaren 2017, 29–30). In the academy, such a lens is necessary to fostering a curiosity beyond the Eurocentricity of the academy, allowing us to understand more clearly what is being done to us in the name of diversity nationally and how current processes of inclusion are linked to the history of exploitation of non-Western societies and complicit in neocolonial interventions here and abroad which work to "impoverish our political imaginations" about how to, in fact, achieve justice (Keating 2012, 226). Within an academic setting then, a decolonial Black feminist political reading practice holds the possibilities for decentering the Western university's investment in a performance of inclusion - as Cory states, we don't have to tell them they are ugly, they already know.

To be clear, I am wary about how the term decolonial has been bantered around the academy as the latest buzzword (Hundle 2019, 298) coopted by the academic elite and in other circles to prove its own progressive and metaphorical (Tuck and Yang 2012) stance through, for example, ceremonial pronouncements of "we are standing on stolen lands," increasingly becoming emptied of meaning and used in the service of "neoliberal market logics" (Hundle 2019, 298). As such, I make it clear in this text that what I am seeking to do is grounded in what Anneeth Kaur Hundle writes as engaging with "[d]ecolonization, in ...[a] more expansive sense" that is working toward "the ongoing undoing of colonization" by interrogating "intellectual genealogies, historical processes, or complex formations of power and subjectivity" in the academy (2019, 298). Like Hundle, I advocate for carefully engaging with anticolonial thought, histories, and practices as we speak back to and challenge the university's current colonial posture and politics with a view to real change (2019, 298) as part of my reading of the academy.

This interrogating includes examining what Jodi Dean calls "the *how* of politics, the ways concepts and issues come to be political common sense and the processes through which locations and populations are rendered as in need of intervention, regulation, or quarantine" (2000, 6), how "the experiences of historically marginalized and minoritized people are constitutive and central to the knowledge formations of the university" (Hundle 2019, 300) which has literally and figuratively deselected, dissected and stitched together their bodies on the conveyor belt of the knowledge production/ assembly line. This focus on the how of politics motivates my questions for this project and the need to examine the ways in which inclusion of the minoritized and marginalized has come to be seen as an issue that can be "taken out of political circulation or are blocked from the agenda" (Dean 2000, 4) because we presume that by including these bodies in the academic space the issues of pervasive discrimination and inequities in the university have been solved. Therefore, how do we understand the how of politics as resulting in the de/politicization, the removal of education equity and justice from the realm of the political, how this removal is facilitated by the practice of inclusion as etiquette and the politics of indifference, and what are its effects?

A decolonial Black feminist reading practice therefore helps us to rethink current diversity and inclusion practices and performances by doing three things. One, it engages with the written histories and archival information within and about the university to expose how its historical legacy of exclusion and exploitation is linked to its current inclusion practices while also moving beyond these histories and archives to engage and privilege the experiences that remain missing, silent, and invisible within these histories and archives. To engage the experiences and memories of those who endure the institution's politics of indifference. This particular reading does what Saidiya Hartmann describes as "to 'brush history against the grain'" (1997, 11). This is, according to Hartmann, to

> excavat[e] at the margins of monumental history in order that the ruins of the dismembered past be retrieved, turning to forms of knowledge and practice not generally considered legitimate objects of historical inquiry or appropriate or adequate sources for history making and attending to the cultivated silence, exclusion, relations of violence and domination that engender the official accounts. [To] struggle within and against the constraints and silences imposed by the nature of the archive – the system that governs the appearance of statements and generates social meaning (1997, 11).

This decolonial Black feminist reading practice therefore seeks to incorporate "the lived experience and discursive production ... [of] voices previously muffled by the loud bass "theory" (Allen 2003,198). It enters into and moves past "the world of fiction" that is the archives (Walcott 2003, 8).

Two, a decolonial Black feminist reading practice of the university engages with the pleasures and disappointments of those students who through Western modernity were/are read as strangers and "space invaders" (Puwar 2004, 8) – women, so-called first-generation, low income, queer, immigrant, Black and Brown people. It is also reflexive, as I pay attention to how I read, that is what emotions do I bring to this reading as a so-called first-generation, immigrant woman from a rural low-income working-class family cum assistant professor at a PWI who is also like these students caught in the crosshairs of the politics of indifference, forced through the apathetic methodology of power to submit to measurement for the purpose of extracting data for university use. Our marked bodies become literal data, they exist *as* and *in* difference, that is, marked as different while navigating the space of difference (read as a space that is good or even exceptional at diversity) as the body that marks that space as such. A decolonial Black feminist reading practice is interested in how these students many of whom are happy/elated/pleased to enter the university, believing in what I call the promise and politics of yearning, exists as difference within difference (read diversity).

By the promise and politics of yearning, I mean the ways in which wanting a better life and future for self, family and community for the minoritized and marginalized has become and been so politicized to the point that modern neoliberal institutions are held out/promised as the sole spaces where we can find, and have fulfillment for these historical and ancestral yearnings that come from deep down in our chest. We are promised that if we want it bad enough and work hard enough, make sacrifices, keep our chin up, that our desires for a better future will be met, and so we remain and endure whatever discrimination is meted out by these institutions which is the price we learn we have to pay. We must believe we cannot give up, and if we fail, it is because we didn't work hard enough or want it bad enough. What I am calling the promise and politics of yearning is similar to Lauren Berlant's concept of cruel optimism. Berlant writes, "optimism is cruel when the object/scene that ignites a sense of possibility actually makes it impossible to attain the expansive transformation for which a person or a people risks striving; and, doubly, it is cruel insofar as the very pleasures of being inside a relation have become sustaining regardless of the content of the relation, such that a person or a world finds itself bound to a situation of profound threat that is, at the same time, profoundly confirming" (2011, 2). In short, "cruel optimism is the condition of maintaining an attachment to a significantly problematic object" (24). As more women and folks of color enter the academy, tuition continues to rise and funding decreases, folks are ending up more and more in debt as they try hard to adapt to and assimilate within the university, to shake off the image of stranger and space invader and to demonstrate that they are worthy to be in this space and to jettison any traits that could make them appear otherwise. They come to embody the promise of yearning as mitigated through the how of the politics of indifference within the university, premised on being recognized through its spatial organization.

And yet there are ways in which this promise simultaneously produces joy stemming from a sense of one's achievement, pride in one's ability to persevere, and anxiety, panic over how much longer one can really endure, hold on, and over the loss of identity. What the promise and politics of yearning makes clear is that despite a discursive commitment to equity "[o]ur society enacts the oppression of cultural imperialism to a large degree through feelings and reactions, and in that respect, oppression is beyond the reach of law and policy to remedy" (Young 1990, 124) because of how deeply it is felt by those who are oppressed. The university generates a particular affectivity in and on marginalized and minoritized bodies such that we feel the university in very specific ways. In fact, the university's historical and political "claim of modern reason to universality and neutrality, and its opposition to affectivity and the body, [has] lead ... to the devaluation and exclusion of some groups" (Young 1990, 124) even as it ensures an affective relationship between itself and those groups. In this way, yearning becomes contested terrain and hence politicized as it is experienced through the promise of transforming those who yearn from affective to rational beings to achieve the better life for which they have yearned. Black feminist poet and writer Audre Lorde has stated that we cannot think about historical conditions produced by European colonialization in simplistic binaristic terms, rather "[i]n a society where the good is defined in terms of profit rather than in terms of human need, there must always be some group of people who, through systematized oppression, *can be made to feel* surplus, to occupy the place of the dehumanized inferior. Within this society, that group is made up of Black and Third World people, working-class people, older people, and Women" [emphasis mine] (1984, 114). Bearing this in mind, I ask, how do these bodies, marked as the bodies that provide the university with much-coveted difference, navigate university difference/diversity as relegated/regulated mostly or solely as that difference, confined to participate in the ways that highlight them as this difference? How do those marked in indifference – the politics and the perception - navigate these spaces *in* difference by conforming to and/or pushing back against said in/difference. I am therefore interested in the negotiation that occurs between universities as they engage (or not) with the politics of indifference, and differenced-bodies in difference, and the consequences of this negotiation, the pleasures and the disappointments.

The thing about Black feminism is that it, according to Nash, is "not simply ... an intellectual, political, creative, and erotic tradition but also ... a way of feeling [and] [t]he felt life of black feminism is varied and complex" (2019, 28). Black feminisms as a way of feeling is "rooted in the intellectual tradition that has voiced the ecstasies, frustrations, longings, and fatigue of scholars" (Nash 2019, 28) whose work has helped me to not only understand theoretically but also relationally why and how people, including me, yearn for better but how they/we activate that yearning, why we take the paths we do to try to fulfil that yearning even through disappointment, hardship and continued oppression, and how that yearning becomes politicized and used against us by the institutions which we seek to help us fulfill it. Black feminist cultural critic bell hooks writes, "people who are writing about domination and oppression are distanced from the pain, the woundedness, the ugliness... [but] resistance begins with people confronting pain, whether it's theirs or somebody else's, and wanting to do something to change it.... Pain as a catalyst for change, for working to change" (1990, 215). Black feminisms then allows me to engage on an affective level as I also pay attention to how as a Black feminist the topics I bring to the fore here can be felt toward working toward change. I ask, how does it *feel* to experience oppression and vulnerability (Nash 2019, 118)? How does it feel to be "included"? In asking these questions, I pay attention to what McKittrick writes in her essay in *Small Axe*, "Rebellion/Invention/Groove" about the ways in which antiblackness informs the "neurobiological and physiological drives, desires, and emotions—and negative feelings" (2016, 121). Acknowledging these feelings as well as how my theorizing is/can be felt by those who I write about, and for, and how I feel doing this work is also critical to this research.

Third and finally, a decolonial Black feminist reading practice of the academy, by exposing the flaws of using an apathetic methodology of power, in the words of Cory, advances to shade. In doing so, it exposes the colonial underpinnings of current inclusion performances and promotes disengagement with such performances by those who serve as the university's difference. This Black feminist shad(e)y theoretics as I call it, demonstrates how the academy continues to marginalize those it seeks to include, as well as non-European knowledges and political traditions while at the same time masking the West's role in undermining those traditions (Shohat and Stam 1994, 15-16). Just like Nance mentioned above engages with a reading of the academy, albeit racists, ableist, and heterosexist, toward making a call to action to a specific group which she refers to as "us," a decolonial Black feminist reading practice of the academy, also calls to "an us," those interested in education justice and in dismantling the ways in which the neoliberal university operates to exploit those who it seeks to include, who are interested in (re)engaging with decolonial ways of doing, knowing and being.

To achieve these three goals, I engage with what Carole Boyce Davies refers to as an "elsewhere" (1994), and what Walcott calls a "whatever" (2003, 121). Elsewhere, Davies writes "denotes movements," the assertion of agency as we cross borders, make journeys, migrate and in so doing re-claim and re-assert (1994, 37). As mentioned earlier, in McKittrick's theorizing on Black geographies, space is created through concealment and subjectivity. However, as Davies makes clear, we can contest this concealment through the creation of an elsewhere which is embedded in transnational and diaspora formulation (1994, 13). For me, what this means is that even as we occupy the t/here of the university we can pull from the different places and spaces and times that have formed us. According to Achille Mbembe, "[d] ecolonizing the university starts with...the rearrangement of spatial relations" (2015, 5). He also writes about the "time of decolonization" that is "the time of closure as well as the time of possibility" (13). As we navigate these oppressive spaces in the current time, our experiences and knowledges gained from and about our elsewhere, including or interior (Quashie 2012), is what allows us to survive and what has the capacity to change. Ahmed writes, "[w]e have a *temporal* movement from the now to the not yet" (2000, 145). How are our temporal and spatial elsewheres bound up in our here and now, how do they inform our yearnings, how can we pull from them and the possibilities they hold out? Who and what do we bring with us from our elsewheres that can guide us through our present? Who and what has made our present spatial occupation possible? What ancestral knowledge, what diaspora connection, family member, childhood experiences, cultural traditions, music, art, stories should we remember that prepared us for the hereness of the knowledge gained within the university?

> We could ask, not only [who and] what made this [present] encounter possible (its historicity), but also what does it make possible, what futures might it open up? At the same time, we have a spatial movement from here to there. We need to ask, not only how did we arrive here, at this particular place, but how is this arrival linked to other places, to an elsewhere that is not simply absent or present? We also need to consider how the *here-ness* of this encounter might affect *where we might yet be going* (Ahmed 2000, 145).

Édouard Glissant writes about the "the union between elsewhere and possibility" (1997, 37-38). It is this union that has the ability to transform how we engage with our hereness of the university, to see beyond it, to pull from our everywheres to imagine and create, construct spaces and knowledges that do not conceal but reveal the possibilities that Black geographies and diasporas can offer.

In theorizing "whatever," Walcott invokes "a transnational sensibility, granted sometimes ephemeral, but always politicized, one that recognizes that *one locality ... cannot stand in for all*" (2003, 121). I invoke Walcott's whatever as a means of wanting to move beyond simply referencing the academy as colonial but recognizing that the academy as a nation-building project is "always already transnationally configured and inflected" (Walcott 2003, 121). Therefore to imagine possibilities for those who exist in difference in the academy requires more than a linear historiography of the space but requires that we look to generating political alternatives (as in whatever), beyond (as in elsewhere) the academy.

Working through Walcott's whatever and Davies elsewhere I return to Saidiya Hartman's work on the obscenity of the display of Black bodies tortured and beaten, where she writes, "rather than try to convey the routinized violence of slavery and its aftermath through invocations of the shocking and the terrible, I have chosen to look elsewhere and consider those scenes in which terror can hardly be discerned … By defamiliarizing the familiar, I hope to illuminate the terror of the mundane and quotidian rather than exploit the shocking spectacle. What concerns me here is the rubric of pleasure, paternalism, and property" (1997, 4). Defamiliarizing the familiar also requires engaging in what she calls "critical fabulation" as I ask, what will people read about this contemporary period decades from now, and reflect on the implications of my own and other's academic/scientific gaze of indifference for the continued gendering and racing of the institution through how I and others write the marginalized and minoritized into academic discourse and practice, not as spectacles but as bodies that think and feel? How does my one voice and one positionality attempt to stand in for others? Who is being eclipsed, who is missing? How do my reading practices work to connect several geographic and temporal localities by examining the legacy of colonial social constructions of the body and the normal in/at the university as means of excluding Others, and the contemporary university's practices of diversity, and how do some connections get lost or neglected? By thinking through whose voices have been historically omitted, I work to make visible a common thread that highlights how the legacy of oppression has managed through the practice of the politics of indifference to achieve the means of upholding the Eurocentric academy (Weheliye 2014) through the omission of histories, voices, experiences, theorizing and ideologies of racialized and sexualized others.

To get to that elsewhere and that whatever of the academy, I engage the theorizing of Black feminist geographer Kathrine McKittrick. McKittrick writes about the *where* of Black feminisms or Black geographies which are located within and outside the boundaries of traditional spaces and places. According to McKittrick

> they expose the limitations of transparent space … [and] locate and speak back to the geographies of modernity, transatlantic slavery, and colonialism; they illustrate the ways in which the raced, classed, gendered, and sexual body is often an indicator of spatial options and the ways in which geography can indicate racialized habitation patterns; they are places and spaces of social, economic, and political denial and resistance (2006, 7).

Like McKittrick, hooks also makes clear that spaces real or imagined "tell stories and unfold histories. Spaces can be interrupted, appropriated and transformed through artistic and literary practice" (1990, 152). It is by thinking and working through the complexity of transparent space of the academic geography that has and continues to deny those whom it has historically excluded and now purports to include as fully human, that we find ways which reveal how ideas get turned into lived and imaginary spaces that are tied to the geographic organization. Black geographies make clear how practices of domination sustained by a unitary vantage point – i.e., those in the university – naturalize both identity and place, repetitively spatializing where and when nondominant groups "naturally" belong. In her theorizing, McKittrick engages with Édouard Glissant's concept, poetics of landscape, which she writes "creates a way to enter into, and challenge, traditional geographic formulations without the familiar tools of maps, charts, official records, and figures; he enters, through his voice-language, a poetic-politics, and conceptualizes his surroundings as 'uncharted,' and inextricably connected to his selfhood and a local community history" (2006, xxii). McKittrick also writes about a "plantation logic" similar to slavery which exists today at both the ideological and material levels and which through negotiations of time, space and violence have emerged different ways of survival through a creolization process that includes the creation of blues music and other expressions (2013, 3). In this same vein, I, like Pratibha Parma, believe that the ways in which we use space are political acts (cited in hooks 1990, 152), and as such, I call for a reimagining of the university as a site "through which particular forces of empire (oppression/resistance, black immortality, racial violence, urbicide) bring forth a poetics that envisions a decolonial future" (McKittrick 2013, 5) and call for a creolization of the university.

What follows in this text then is a decolonial Black feminist reading practice of the academy, a shad(e)y theoretics, where I demonstrate how as Black feminists, we should be invested in the spatial politics of the academy precisely because we have been relegated to the margins of knowledge and have therefore been imagined as outside of the production of this space even as the exploitation of our bodies, and our labor has been and continues to be integral to its building and success. This project is invested in making space for new imaginings of what we perceive as knowledge and ways of being in relation. In this reimagining, we must insist that black feminist politics is a location from where we speak and disrupt the diversity and inclusion metanarrative. Utilizing this lens helps me to go beyond the obviousness of difference to understand how it is, in fact, the politics of indifference that drives the practice of diversity and inclusion as we know and have experienced it.

In Chapter 1 of this text, I engage in a historiography of the contemporary university to show how it is the product of colonial mappings. I explain how the use of science and data originating in the university was used to justify its own practices of exclusion as well as to provide evidence for oppression in the larger society by constructing some bodies as other than human and as nonpersons. In Chapter 2, I explain in more detail what I mean by the politics of indifference and the apathetic methodology of power, how they are produced but also produce data about the body that work to negatively impact the marginalized and minoritized. I also juxtapose the term politics of indifference used to frame this project with Iris Marion Young's politics of (positional) difference to demonstrate that while the politics of (positional) difference "worry about the domination some groups are able to exercise over public meaning in ways that limit the freedom or curtail opportunity... [and] challenge difference-blind public principle" (2005, 28), the politics of indifferences goes deeper to examine what is behind this domination, that is what has been concealed, or the how of the politics of this domination, that is, the ways in which it is intentionally wielded, enacted and propagated through very specific and methodological uses of power. This chapter demonstrates how the politics of indifference requires turning a blind eye to the consequences of practices and policies which it causes to be implemented (including its exclusion/inclusion policies – inclusive orientations, pronouncements of solidarity, using pronouns) and which can result in incidents of surveillance, trauma, and even death. For example, we know that there is a problem with sexual assaults of women on college campuses (Rainn) and while campuses tout the number of women in their diversity statistics, what are they doing to seriously address issues of sexual assault on campus? Another example is "how intersectionality has been rhetorically and symbolically collapsed into diversity, and this taken up as an inclusion project that resonate[s] with the mission of the so-called corporate university" (Nash 2019, 11-12) touting the recruitment numbers of students who make up their diversity while consistently failing to retain Black and Brown students who continue to be and feel marginalized. I also use this chapter to define other terms mentioned briefly in this introduction such as, the promise and politics of yearning, engaging with affect theory to demonstrate how the affective plays a crucial role in the ways in which certain bodies experience feelings of being included/excluded and how these feelings stem from a history of how said bodies have experienced inclusion/exclusion. I use the remainder of the chapter to demonstrate how a decolonial Black feminist reading practice allows us to examine difference in contemporary societies. I ask questions such as, does the university today, through the extension of a sense of belonging "exemplify the use of the body as an instrument against the self" and hence rewrite on that body its originally subjugated status (Hartman 1997, 22)? How does the fascination and satisfaction of being seen as benevolent/altruistic in the academy as inclusive go hand in hand with the continued subjugation of the marginalized and minoritized?

I use the following chapter to provide examples of how the politics of indifference works and demonstrate how inclusion and exclusion are indeed two sides of the same coin. Chapter 3 focuses mainly on the sole narrative of a Black woman student whose experience I use to demonstrate the deeply affective impacts of being the university's diversity. Using Christina Sharpe's theorizing of the weather, I posit that in the academy where, according to Jennifer L. Morgan, "the intellectual work necessary to naturalize African plantation enslavement – that is, the development of racialist discourse – was deeply implicated by gendered notions of difference and human hierarchy," (1997, 170) and "mutually constitutive ideologies of race and gender"

(170), the narrative of this Black woman, who I call Clare, is crucial to understanding how the formation of the colonial world, which the university validated through the measuring of black people's skulls, the theft of black peoples dead bodies, the measuring of college students to prove normality, morality, intelligence and fitness still lives on. These measurements are not just things that happened in the past, and the academy should be seen as part of what Lisa Lowe calls "[t]he operations that pronounce colonial divisions of humanity ... imbricated processes, not sequential events ... ongoing and continuous in our contemporary moment, not temporally distinct nor as yet concluded" (2015, 7).

Acknowledging these linkages help us to turn to the necessary work of "'radical delinking' from the continuities of inequalities and inequities that are being repeated through neoliberalism" (Sultana 2019, 35) which I turn to in the next chapter. In Chapter 4, I engage with the concepts of elsewhere and whatever to *read* the university toward what Sharpe calls aspiration, or to put life back into the black body (2016), thingified (through data) by the colonialist legacy of the university. Sharpe's concepts of aspiration and wake work are important, as I shift from reading to shade and toward elsewhere possibilities, engaging with concepts of relationality (Alexander and Mohanty 2012, xix) and to, as Sylvia Wynter puts it, "'being human as praxis' and think outside the logic of coordinated exclusion and profitable conformity" (McKittrick 2015, 154). In *Black Women, Writing and Identity: Migrations of the Subject* Carole Boyce Davies writes reflexively about Caribbean children creating elsewhere. She writes,

> [t]his writer is also of that generation who as children witnessed beginnings, grew up with the language of independence and new flags, anthems, etc., celebrated and them promptly recognized as adults that there were new colonial formations and American imperialism and how painful it had been to be put out in the sun with flags as children to wave at some member of the royal family as we marched to independence. So for us it is necessary to create "elsewhere" worlds and places and consciousness (1994, 89).

While I was born more than a decade after independence, the legacy of a colonialist education system limited my imagination of elsewhere spaces. I remember learning about imperialist peoples and places in my required readings on British and U.S. literature and history and remember learning to desire them. I think about how as a young girl attending one of the most prestigious secondary schools in my country, I longed to do well enough to get a scholarship to go to Warwick University in England to study law. I didn't know anything about Warwick but I requested an information packet and when I received it in the mail, I longed for this place even more. What is not lost on me now as an assistant professor working at a research one institution in the United States, is how these longings are learned and produced through centuries of inequities and displacements across civilizations, which not only teaches us that our cultures are not good enough, but also create conditions that produce hardships as almost a self-fulfilling prophesy. I also think about how desiring bodies like mine are recruited by neoliberal universities in the West to enhance their appearance of diversity, our bodies used as data to show they have X number of international scholars comparable to peer institutions. I remember when I first started teaching in the US academy, though a proud graduate of the Institute for Gender and Development Studies Nita Barrow Unit at the University of the West Indies (UWI) Cave Hill campus, I would often be very slow to acknowledge where I was educated when my colleagues would ask while mentioning they went to Brown, or Harvard, or Yale; a manifestation of this residual colonial longing for Warwick. And yet, my UWI experience at an Institute named after a Black Caribbean feminist pioneer, and the camaraderie experienced among Black Caribbean women was critical to my growth as a feminist scholar. As I now teach in a place where Black feminist courses are sparse and where I am the only Black woman in the Women's and Gender Studies program, I long to return to this other, elsewhere place and time. As I teach and work in this place I once desired and where my body now sits as data indicating that the university has a Black feminist on campus who does XYZ, I long for an elsewhere and a whatever.

I reflect on Davies work, as she writes about the work of Claudette Williams in "Gal ... You Come From Foreign" (2002, 103), where Williams "reflects on the difficulties and pleasures of life both at home in the Caribbean and in London and writes "I still possess a strong emotional attachment to the concept of "back home"; England has never emotionally become my home, even though I've lived here some twenty years now" (51). Similarly, this t/here-ness for me is not simply about my arrival and existence in the academy as it currently operates or a longing for a past that is unattainable and a back home that only now exists in my memory, but (a) place(s) from which I can pull as I take up my habitation in the where of Black feminism. Pulling from these spaces helps me to hold out hope about how the work I am able to do here, in this present, as it has been impacted by temporal and spatial elsewheres, might impact other spaces and times.

The concluding chapter of this text engages with a concept of creolization and advances a creolization of the university. Building on Édouard Glissants' *Poetics of Relation* (1997) and Edward Kamau Brathwaite's theorizing about creolization (1971) to demonstrate that on the spectrum of staying fixedly in place at the university on the one end, and burning it all down on the other, there are ways to imagine and achieve a decolonial elsewhere that is not beholden to transparent space, but that is capable of doing the hard decolonial work of excavating to produce new modes of being human. This final chapter embraces Glissant's concept of the poetics of relation wherein he describes, "two conditions ....com[ing] together ... Under these conditions poetic thought went on the alert: beneath the fantasy of domination it sought the really livable world" (1997, 28). Beneath the fantasy of domination and all the ways the academy would build and be built by this domination, there is this poetics one that has the possibility to metamorphosize, to produce "cognitive schemas, modes of being human that ... restructure our existing system of knowledge" (McKittrick 2016, 81).

In his 1999 seminal text *Disidentifications: Queers of Color and the Performance of Politics,* queer theorist José Esteban Muñoz writes about "the burden of liveness" which he describes as "a particular hegemonic mandate that calls the minoritarian subject to 'be live' for the purpose of entertaining elites. This 'burden of liveness' is a cultural imperative within the majoritarian public sphere that denies subalterns access to larger channels of representation, while calling the minoritarian subject to the stage, performing her or his alterity as a consumable local spectacle" (182). When I think about Muñoz's burden of liveness, I am reminded of Hartman's work quoted earlier on consenting agency and how in the academy, those "included" are forced to perform their difference for the benefit of the academy, such that difference consigns and confines how they can participate, denying them the ability to show up as anything other than difference. Using the example of the Chusmería, which he describes as linked to a stigmatized identity, Muñoz writes "[m]inoritarian subjects do not always dance because they are happy; sometimes they dance because their feet are being shot at" (189). When I think of this statement, I think of the article published in *The Atlantic* in April 2019 entitled "The Death of An Adjunct" (Harris 2019, n.p.). The article told the story of Thea Hunter, a Black woman adjunct who did work she loved but whose feet were being shot at as she was forced to perform as someone included but "exploited by a system that consumes" (Harris 2019, n.p.). Thea is dead and the system that exploited her, and continues to exploit us remains intact. As most of us continue to move to the sound of ricocheting bullets and reverberation around our feet, our timed movements we hope will promise the better life we yearn for if we just keep moving and performing our alterity at the behest of the university for which we think we are lucky to be a part.

But when I think of Muñoz's statement, I also think about bell hooks theorizing of marginality also as "a place ... one chooses as site of resistance - as location of radical openness and possibility" (1990, 153). Muñoz's statement was not that we always dance because our feet are being shot at, but rather only sometimes. He leaves open the possibility that there are times where the minoritarian dances because even though they are oppressed, they can, in fact, as agentic beings dance because they are truly happy. According to Chinn, "[b]odies are not simply acted upon, though. We are not just moved through space by ideology, or discourse, or culture, or whatever name one might give the structures of intelligibility, power, and feeling that makes sense of the world for us ... Even within a system of subordination, human relations are active and dynamic: those acted upon also act, in however limited a way" (2000, 21). They can operate from that/those space/s of marginality that is/are a site/s of resistance, that elsewhere, that according to hooks "is continually formed in that segregated culture of opposition that is our critical response to domination. We come to this space through suffering and pain, through struggle" (1990, 153). We come undone by it, we cry, we wail, we beat desks and cuss. But in this undoing, in this knowledge of the struggle as hard, we also "know struggle to be that which pleasures, delights, and fulfills desire. We are transformed, individually, collectively, as we make radical creative space which affirms and sustains our subjectivity, which gives us a new" (hooks 1990, 153) reason to dance and sing. This dancing and singing is similar to Glissant poetics mentioned earlier and to Wynter, quoted in McKittrick as stating that attention needs to be drawn "to how the creation of culture, the making and praxis of music—within the context of hateful and violent ... axioms—is underwritten by 'the revolutionary demand for happiness'" (2016, 81). This type of dancing stems from a place of recognizing our vulnerability to being killed when our feet are shot at, recognizing our vulnerability by being invited to dance/perform for the majoritarian. But if we are to dance/perform, we can also witnesses, witness that "we are subjected to violence, particularly by social structures that have been constructed to discipline and surveil" (Nash 2019, 119), and therefore plan. Centering this vulnerability and witnessing is the work that this text, like many other Black feminist works, according to Nash, has invested in as love as political practice (2019, 121). Using this political practice to plan, much like Stefano Harney and Fred Moten write about the undercommons "is not an activity, not fishing or dancing or teaching or loving, but the ceaseless experiment with the futurial presence of the forms of life that make such activities possible" (2013, 74–75).

I also chose to reference Muñoz here because he articulates in his work how to make this future possible through the concept of worldmaking. For him, worldmaking is "the ways in which performances ...have the ability to establish alternate views of the world. These alternative vistas are more than simply views or perspectives; they are oppositional ideologies that function as critiques of oppressive regimes of 'truth' that subjugate minoritarian people" (1999, 195). In this decolonial Black feminist reading practice of the university, I give examples of these types of performances as examples of alternative vistas that seek not only to critique the oppressive truth claims and resulting politics and practices stemming therefrom but also to propose a new world, an elsewhere of the academy. Black feminist Audre Lorde wrote in 1988 that "[w]ithout a rigorous and consistent evaluation of what kind of a future we wish to create, and a scrupulous examination of the expressions of power we choose to incorporate into all our relationships including our most private ones, we are not progressing, but merely recasting our own characters in the same old weary drama" (1988, 1). Lorde's words regarding the types of curation we must engage to ensure our future remains ever-present. As I write this text, I also come back to a question Katherine McKittrick asks, "[h]ow then do we think and write and share as decolonial scholars and foster a commitment to acknowledging violence and undoing its persistent frame, rather than simply analytically reprising violence" (2014, 18)?

This is my purpose for this project, particularly at a time when there has been a resurgence of overtly bigoted white nationalism in the United States, so much so that even problematic university diversity and inclusion initiative are called into question by people like Nance (Hundle 2019, 296). In this time, where the university pretends to value the lives and bodies of the minoritized and marginalized, putting out statements about Black lives matter and the racial uprisings after the deaths of George Floyd and Breonna Taylor in solidarity, and yet are complicit in the Trump administration's overtly bigoted actions to end diversity trainings and other actions that devalue said lives, it is important that we engage the elsewhere and the whatever. We cannot value ourselves based on the university's metric systems that counts and recounts us for its own futurity. Daina Ramey Berry, in her 2017 text *The Price For Their Pound of Flesh: The Value of the Enslaved, from Womb to Grave, in the Building of a Nation*, a text that addresses the different and varied ways the Black enslaved were valued in life and death, theorizes value as "a noun, a verb, and an adjective ... active, passive, subjected, and reflexive ... 'rooted in modes or kinds of *valuing*' and requires an assessment of feelings" (6). Defining value in this way is important as we think about the (else)where and the what(ever) of ourselves. As Berry states in her book, the enslaved had what she calls "*spirit* or *soul* value" "an internal quality ... an intangible marker that often defied monetization ...spoke to the spirit and soul of who they were as human beings ... represented the self-worth of enslaved people" (6). Soul values are based on an inner centering, much like Lorde writes about the erotic and Quashie the interior and the sovereignty of quiet. It was reinforced by communities and for the enslaved sometimes "appeared as a spirit, a voice, a vision, a premonition, a sermon, an ancestor, (a) God. It came in public and private settings it was occasionally described as a personal message from a higher being, a heaviness in the core of their bodies. ... Soul values ... came from deep within a person's heart" (61) it is what moved their feet to dance otherwise.

As the university, just as the dominant and transparent geography of the plantation, fixed value (market, sale price, diversity and inclusion), based on computations, assessments, appraisals, policies, and reports on, and about laboring black flesh, our quantified bodies possess/ed an erotic/soul value/ interior for which "outsiders did not bargain for ... [and that] shaped and defined" (Berry 2017, 61) us. In mentioning the above, however, we also need to acknowledge what Patricia J. Williams calls "spirit murder" (1992) that is a "psychic and spiritual wounding ... left on the flesh, psyche, and even soul of those who experience violence *and* the wounds, often invisible, that haunt perpetrators of violence, including a willingness to accept, and to render *unseen*, those who are dispossessed" (Nash 2019, 123–124). Being vigilant for and about the unseen, the nontransparent through engaging in the elsewhere and the whatever then is crucial to saving our souls.

References

Ahmed, Sara. *Strange Encounters: Embodies Others in Post-Coloniality*. London: Routledge, 2000.

Ahmed, Sara. *On Being Included: Racism and Diversity in Institutional Life*. Durham: Duke University Press, 2012.

Alexander, M. Jacqui and Chandra Talpade Mohanty. "Introduction: Genealogies, Legacies, Movements." *In Feminist Genealogies, Colonial Legacies, Democratic Futures*, edited by M. Jacqui Alexander and Chandra Talpade Mohanty, xiii–xlii. New York: Routledge, 2012.

Allen, Jafari Sinclaire. "Black Diaspora Genealogies from 'Niggernicity' to Manifold Futures." In *Decolonizing the Academy: African Diaspora Studies*, edited by Carole Boyce Davies, Meredith Gadsby, Charles Peterson and Henrietta Williams, 171–204. Trenton: Africa World Press Inc, 2003.

Berlant, Lauren. *Cruel Optimism*. Durham: Duke University Press, 2011.

Berry, Daina R. *The Price for Their Pound of Flesh: The Value of the Enslaved, from Womb to Grave, in the Building of a Nation*. Boston: Beacon Press, 2017.

Brathwaite, Edward Kamau. *The Development of Creole Society in Jamaica: 1770–1820*. Oxford: Clarendon Press, 1971.

"Campus Sexual Violence: Statistics," Rainn, https://www.rainn.org/statistics/campus-sexual-violence.

Celikates, Robin and Yolande Jansen. "Reclaiming Democracy: An Interview with Wendy Brown on Occupy Sovereignty, and Secularism," *Critical Legal Thinking*, January 30, 2013, https://criticallegalthinking.com/2013/01/30/reclaiming-democracy-an-interview-with-wendy-brown-on-occupy-sovereignty-and-secularism/.

Chinn, Sarah E. *Technology and the Logic of American Racism: A Cultural History of the Body as Evidence*. London: Continuum, 2000.

Davies, Carole Boyce. *Black Women, Writing and Identity: Migrations of the Subject*. New York: Routledge, 1994.

Davis, Angela. *Freedom Is a Constant Struggle: Ferguson, Palestine, and the Foundations of a Movement*. Chicago: Haymarket Books, 2016.

Dean, Jodi. "Introduction: The Interface of Political Theory and Cultural Studies." In *Cultural Studies and Political Theory*, edited by Jodi Dean, 1–20. Ithaca, NY: Cornell University Press, 2000.

"Economic Impact of International Students," *iie*, https://www.iie.org/Research-and-Insights/Open-Doors/Economic-Impact-of-International-Students.

Ella, Shohat and Robert Stam. *Unthinking Eurocentrism*. New York: Routledge, 1994.

Glissant, Édouard. *Poetics of Relation*. Ann Arbor: University of Michigan Press, 1997.

Hall, Stuart. "Part 2: Thatcherism: Racism and Reaction [written in 1978]." In *Political Writings: The Great Moving Right Show and Other Essays*, edited by Sally Davison, David Featherstone, Michael Rustin and Bill Schwarz, 142–157. Durham: Duke University Press, 2017.

Harney, Stefano and Fred Moten. *The Undercommons: Fugitive Planning & Black Study*. New York: Minor Compositions, 2013.

Harris, Adam "The Death of an Adjunct." *The Atlantic* (2019).

Hartman, Saidiya V. *Scenes of Subjection: Terror, Slavery, and Self-Making in Nineteenth-Century America*. New York: Oxford University Press, 1997.

hooks, bell. *Yearning: Race, Gender, and Cultural Politics.* Boston: South End Press, 1990.

Hundle, Anneeth Kaur. "Decolonizing Diversity: The Transnational Politics of Minority Racial Difference." *Public Culture* Vol. 31, no. 2 (2019): 289–322. doi: 10.1215/08992363-7286837.

Keating, Christine. "Against the Politics of Compensatory Domination." In *Feminist Solidarity at the Crossroads: Intersectional Women's Studies for Transracial Alliance,* edited by Kim Marie Vaz and Gary L. Lemons, 220–227. New York: Routledge, 2012.

Livingston, Jennie. *Paris Is Burning,* Off-White Productions, 1991.

Lorde, Audre. "Age, Race, Class and Sex: Women Redefining Difference." In *Sister Outsider,* 114–123. Berkeley, CA: Crossing Press, 1984.

Lorde, Audre. *A Burst of Light: Essays.* New York: Firebrand Books, 1988.

Lowe, Lisa. *The Intimacies of Four Continents.* Durham: Duke University Press, 2015.

Maldonado-Torres, Nelson. "On the Coloniality of Being: Contributions to the Development of a Concept." *Cultural Studies* Vol. 21, no. 2–3 (2007): 240–270. doi: 10.1080/09502380601162548.

Mbembe, Achille. *Decolonizing Knowledge and the Question of the Archive.* Africa is a Country, 2015. https://africaisacountry.atavist.com/decolonizing-knowledge-and-the-question-of-the-archive.

McKittrick, Katherine. *Demonic Grounds: Black Women and the Cartographies of Struggle.* Minnesota: University of Minnesota Press, 2006.

McKittrick, Katherine. "Plantation Futures." *Small Axe* Vol. 17, no. 3 (2013): 1–15. https://read.dukeupress.edu/small-axe/article-abstract/17/3%20(42)/1/33296/Plantation-Futures.

McKittrick, Katherine. "Mathematics Black Life." *The Black Scholar* Vol. 44, no. 2 (2014): 16–28.

McKittrick, Katherine. "Axis, Bold as Love: On Sylvia Wynter, Jimi Hendrix, and the Promise of Science." In *Sylvia Wynter: On Being Human as Praxis,* edited by Katherine McKittrick, 142–163. Durham: Duke University Press, 2015.

McKittrick, Katherine. "Rebellion/Invention/Groove." *Small Axe* Vol. 20, no. 1 (2016): 79–91. doi: 10.1215/07990537-3481558.

McLaren, Margaret A. "Introduction: Decolonizing Feminism." In *Decolonizing Feminism Transnational Feminism and Globalization,* edited by Margaret A. McLaren, 21–61. London: Rowan and Littlefield International, 2017.

Morgan, Jennifer L. "'Some Could Suckle over Their Shoulder': Male Travelers, Female Bodies, and the Gendering of Racial Ideology, 1500–1770." *William and Mary Quarterly* Vol. 54, no. 1, *Constructing Race* (1997):s 170. doi: 10.2307/2953316.

Muñoz, José Esteban. *Disidentifications: Queers of Color and the Performance of Politics.* Minnesota: University of Minnesota Press, 1999.

Nance, Penny. "My Son's Freshman Orientation At Virginia Tech Was Full Of Leftist Propaganda." *The Federalist,* August 14, 2019, https://thefederalist.com/2019/08/14/sons-freshman-orientation-virginia-tech-full-leftist-propaganda/.

Nash, Jennifer C. *Black Feminism Reimagined: After Intersectionality.* North Carolina: Duke University Press, 2019.

Puwar, Nirmal. *Space Invaders: Race, Gender and Bodies Out of Place.* Oxford: Berg, 2004.

Quashie, Kevin. *The Sovereignty of Quiet: Beyond Resistance in Black Culture*. New Jersey: Rutgers University Press, 2012.

Sharpe, Christina. *In the Wake: On Blackness and Being*. Durham: Duke University Press, 2016.

Sultana, Farhana. "Decolonizing Development Education and the Pursuit of Social Justice." *Human Geography* Vol. 12, no. 3 (2019): 31–46. doi: 10.1177/194277861901200305.

Thompson, A.C. "Inside the Secret Border Patrol Facebook Group Where Agents Joke About Migrant Deaths and Post Sexist Memes." *ProPublica*, July 1, 2019, https://www.propublica.org/article/secret-border-patrol-facebook-group-agents-joke-about-migrant-deaths-post-sexist-memes?utm_content=buffer58eae&utm_medium=social&utm_source=twitter&utm_campaign=buffer.

"Trump Administration Releases Final Public Charge Rule," American Council on Education, August 12, 2019, https://www.acenet.edu/News-Room/Pages/Trump-Administration-Releases-Final-Public-Charge-Rule.aspx.

Tuck, Eve and K. Wayne Yang, "Decolonization Is Not a Metaphor." *Decolonization: Indigeneity, Education & Society* Vol. 1, no.1 (2012): 1–40.

Walcott, Rinaldo. "Beyond the 'Nation Thing': Black Studies, Cultural Studies, and Diaspora Discourse (or the Post-Black Studies Moment)." In *Decolonizing the Academy: African Diaspora Studies*, edited by Carole Boyce Davies, Meredith Gadsby, Charles Peterson and Henrietta Williams, 107–124. Trenton: Africa World Press Inc, 2003.

Weheliye, Alexander G. *Habeas Viscus: Racializing Assemblages, Biopolitics, and Black Feminist Theories of the Human*. Durham: Duke University Press, 2014.

Williams, Patricia J. *The Alchemy of Race and Rights*. Cambridge: Harvard University Press, 1991.

Young, Iris Marion. *Justice and the Politics of Difference*. Princeton: Princeton University Press, 1990.

Young, Iris Marion. "Structural Injustice and the Politics of Difference." AHRC Centre for Law, Gender, and Sexuality, 2005.

1 Reading the university
Coloniality and the making of the human

In the introduction to the edited text *Decolonizing the Academy: African Diaspora Studies* (Davies, Gadsby, Peterson and Williams 2003), Carole Boyce Davies writes:

> [i]t is intended that the academy is perhaps the most colonized space. By this I mean, it is a site for the production and re-production of a variety of discourses which keep in place certain colonial structures which have as their intent the maintenance of Euro-American hegemonies at the level of thinking and therefore in the larger material world. ... The various disciplines have functioned as sites for the legitimization of Eurocentric knowledge, and are often key locations for the creation and transmission of state ideologies. And, indeed the very philosophical basis on which the academy rests is one which functions to privilege knowledge emanating from European thinkers, to legitimate European belief systems, histories, ideologies, principles, literatures (ix).

It is from this premise that this chapter begins to examine the university and its historical origins, to uncover the ways in which the academy has worked to privilege European knowledges by the concealment and destruction of others through a so-called scientific method which has labeled those whose knowledges they sought to conceal as primitive, mere superstition (Quijano 2008), and irrational. In this chapter, I seek to demonstrate how the long history of both racialized and gendered violence has been operationalized, and continues to show up and impact the ways in which racialized and gendered bodies experience the academy through "attendant knowledge systems that produce this ... violence as 'common sense'" (McKittrick 2005, 2). The importance of examining this history cannot be overstated as the university is a reflection of the nation-building project that is the United States of the past, the present and the future and, according to M. Jacqui Alexander, "[w]e live the privilege of believing the official story ... to consume an education that sanctions the academy's complicity in the exercise of normativization of state terror" (2005, 2).

DOI: 10.4324/9781003019442-2

State terror deployed through institutional power is built into the university's core. Universities in the United States historically and contemporarily have been built upon the backs of the marginalized and the minoritized, and oppression and discrimination are deeply embedded within their infrastructure, with the "founding, financing, and development of higher education [being] thoroughly intertwined with the economic and social forces that transformed West and Central Africa through the slave trade and devastated indigenous nations in the Americas. The academy was a beneficiary and defender of these processes" (Wilder 2013, 1-2) through perpetuation of racist myths as fact (Elliott-Cooper 2018, 292) and the exploitation of enslaved African and Native American peoples labor and land (Meyerhoff 2019, 201). As with racial/ethnic discrimination, gender discrimination and subjection was also, and still is, part of how the university operates. According to Denise Ferreira da Silva, "articulated in the founding statements of modern thought: while the female's role in (physical) reproduction would seem to immediately explain her incarceration in domesticity, gender subjection rests on the liberal rewriting of patriarchy as a juridical-moral moment ruled by 'natural (divine) law'" (2007, xxviii) propagated by and built into the fabric of the university.

But this is not a text written simply to (re)tell the colonial and patriarchal past of the academy, as much more has been done before by scholars and historians better situated to do this telling (see Wilder 2013; Meyerhoff 2019). As a Black woman I am completely aware of how, as Nikol Alexander-Floyd writes, at the foundation of US intellectual culture are racist and sexist assumptions that preclude Black women from the life of the mind (2010, 810). As such, I am much more invested in *reading* the university that constructs me as incapable of producing knowledge than in a retelling of its history. In fact, like Lisa Lowe, I see "[t]he operations that pronounce colonial divisions of humanity" which include the academy, as "imbricated processes, not sequential events … ongoing and continuous in our contemporary moment, not temporally distinct now as yet concluded" (2015, 7). My thesis dives into reading and reading practices, a practice akin to what Eli Meyerhoff calls "Spilling the histories behind these reified ideals [to] destabilize our subscriptions to them" (2019, 204). "Spilling the histories" calls to mind, "Spilling the tea," and is in accordance with the Shad(e)y theoretics which I employ, one that engages Black Feminist, Affect, and Queer theorizing as subversive discursive process. Reading this history helps us "to debunk and expose dominant ideologies such as colorblindness, meritocracy, and equal opportunity … [that] lead people to believe that the educational system is not racist or sexist; failure for women and people of color occurs because students of color and women simply do not work hard enough. As such, the blame for failure is not placed upon systemic inequalities, but individuals who are oppressed by them" (Vaccaro and Cambra-Kelllsay 2016, 50). This reading practice means we need to look beyond the obvious, and to mine and excavate historical connections and thus reconstruct knowledges which connect racist and sexist spatial-izing acts to the present seeming benevolence of the university, in order to bring to the fore how new iterations of dominant geographies within the academy operate. Like Wynter, I am not interested simply in conveying ideas about the university but instead in "the difficult *labor* of thinking ... anew" (McKittrick 2015, 6–7).

As mentioned in the introduction to this text, reading practices, accord-ing to Walcott, "are the histories, memories, desires, free associations, dis-appointments, pleasures and investments we bring to any given texts ... informed by the peregrinations of ... consciousness" (2003, 118). Analyzing these prejudicial university practices through a decolonial Black feminist reading practice then, helps to unearth the various marginalized identity positions occupied by those who, through colonial knowledge production used to justify their oppression and exclusion, are contemporarily con-structed as different, even while the institution's relatively newfound diver-sity initiatives simultaneously tout their inclusion on the basis of that very construction. According to Kelvin Santiago-Valles, reading, "'race,' moder-nity, capitalism, together with chattel slavery and its legacies, are histori-cally and conceptually bound as *coloniality* ... This entire ensemble arose simultaneously five centuries ago as a coupled asymmetric structure and as conflicting hegemonic forms of knowledge and power for (re)ordering time, space, and bodies, a structure which – though transformed – continues to organize the world to this day" (2003, 218). In her work on the construc-tion of Black women as "flesh," Black feminist Hortense Spillers writes of the existence of a "telegraphic coding ... markers so loaded with mythical prepossession that there is no easy way for the agents buried beneath them to come clean" (1997, 65). To make these prepossessions visible requires more than a retelling but rather a good reading to uncover the ways in which academic language and measurements have been used to perpetuate and justify difference through the body in this "historical present ... *situation*" (Berlant 2011, 195). Spillers also writes that as a Black woman, we must "strip down through layers of attenuated meanings, made in excess over time, assigned by a particular historical order" if we are "to speak a truer word concerning ... [ourselves] ... and there await whatever marvels of [our] own inventiveness" (1997, 65). It is the same disappointment that Walcott writes about, but also the marvel, the pleasures that come with the invest-ment in reading that is important to a good read. Even as I write this chap-ter, I am in the same position as you, the reader, as from chapter to chapter, I too am discovering where this read leads.

Important to this reading practice is demonstrating the direct connec-tion between colonization and the creation and circulation of European knowledges from which the university benefits. Recognizing this connec-tion and shifting away from European knowledges is important to the types of methods we create for understanding the problematics of race and gen-der today that are not just an "extension of intersectionality [or]... include another form of difference" (Alcoff 2017, 50). In other words, we need to understand what Alexander refers to as the university's "regressive internal politics" (2005, 13), how these politics are predicated on its outward facing free market liberal ideology and ideas of meritocracy (Alexander 2005, 113), and the ways in which these combine to impact how those experiencing the university today experience feelings of desire, disappointments, and pleasures based on the ways in which they have historically been included/excluded therefrom. The overall structure of the academy is built on neoliberal capitalist foundations steeped in colonialist racism and misogyny, and it is critical to identify how this history still has real-life impacts in relation to systematized discrimination, exclusion, and social reproduction which continue to stick to the bodies of the marginalized and minoritized groups. As such, this reading practice also requires contesting the formations of gender-based and raced-based social organization and the ways in which this organization is globally interconnected and historically supported by the academy.

According to Achille Mbembe, "[t]oday, they are *large systems of authoritative control*, standardization, gradation, accountancy, classification, credits and penalties. We need to decolonize the systems of management insofar as they have turned higher education into a marketable product bought and sold by standard units" (2015, 7). Since its founding the university has been responsible for enacting epistemic violence and sanctioning the physical violence of Black and Brown folks and women through these forms of classification that has produced "the terrible marks of gender and race" as imbricated in each other (Haraway 1989, 1). "Hierarchies of race and gender *require* one another as co-originating and co-dependent forms of oppression" (Doyle 1994, 21) produced through modes of power and buttressed by ways of thinking that continue to influence all aspects of life in capitalist society (Mendoza 2016, 119). According to Grace Kyungwon Hong, "long histories of colonialism, enslavement, and genocide were the ways in which the farthest reaches of the globe became connected in the prior eras of a decidedly global capitalism" (2008, 99). Michael Hanchard writes that while they were capturing and enslaving Africans, uprooting and scattering them all across the New World, Europeans and later Americans were constructing national dialogues justifying these atrocious acts (1990, 40). As the so-called founding fathers of the United States created a system of citizenship they justified the exclusion of a number of people "by imbuing them with a set of attributes that made them unfit for citizenship" including claims that non-white groups were savage, irrational, and lacked the self-control necessary for self-governance (Roberts 2017, 8). These systems of violence came to be cemented through what Hong calls "an epistemology of white supremacy" (2008, 99), which required/s the reinforcement of those inferior designations by organized repression of the non-white/non-male subject's possibility of proving them wrong. This type of violence marked "the status of the nonperson," (Perry 2018, 35) producing a conception that differentiated the world into "two groups: superior and inferior, rational and irrational, primitive and civilized, traditional and modern" (Lugones 2016, 4).

As such, according to Angela Davis, "[f]or most of our history the very category 'human' has not embraced Black and Brown people. Its abstractness has been colored white and gendered male" (2016, 87). Davis' point is important in a book about the university because, as Denise Ferreira da Silva states "[a]mong existing things, humanity is highest in the figuring of determinacy because it alone shares in the determining powers of universal reason, since it alone has free will, or self-determination" (2017, 8). As such, to be considered outside of the realm of the human is to be classified as without reason and therefore incapable of self-determination. In her work on the genre of human, Black feminist philosopher Sylvia Wynter writes about a Chain of Being that represents degrees of rationality among people and civilizations; a chain which the West created and upon it placed themselves at the top and Black civilization and people at the bottom – "at the nadir of its Chain of Being; that is, on a rung of the ladder lower than that of all humans" (Wynter 2003, 300–301). In making this point, Wynter also demonstrates that there are what she calls a whole host of people who do not fit the overrepresentation of the current mode of being human and are thus banished to the archipelago of human otherness, those who are non-white, woman, queer, etc. (300–301).

Others who have theorized the human in this way include Aimé Césaire who points out, "colonization = 'thingification'" (1972, 21); da Silva who writes, "Wynter's reading ... shows that, *beyond* providing the grounds for the abstract mode of comparison (measurement or classification that resolves difference in a glassy text as taxonomy or mathesis), *universal reason*, precisely because it is the ground for the rational/irrational pair, refigures the medieval Spirit/Flesh divide and *sustains* the writing of European particularity" (2015, 97); and Nelson Maldonado-Torres who writes about how the colonizers consciousness structures what he calls the Western *cogito* to refuse to consider the humanity of non-white people (2007). Like Maldonado-Torres, Black feminist Cheryl I. Harris theorizes about whiteness "as racialized privilege ... legitimated by science ...'objective fact'" (1993, 1738), and Iris Marion Young writes about how "modern scientific and philosophical discourse explicitly propound and legitimate formal theories of race, sex, age, and national superiority" (1990, 125).

There have been and continue to be tremendous consequences resulting from the use of so-called universal reason designed by Euro-Americans to classify themselves as human and others as non-human. These, according to Tiffany Lethabo King, "methodologically ... bloody, bodily, discursive, sensual (and effective) enactments of perverse and gratuitous violence ... [were] theoretically approached and narrated as a way of deftly and surgically reading the minutia of its quotidian discursive moves and affectations" (2016, 6). While the consequences are varied, what often goes untheorized regarding the impacts of such classification are the affective consequences. For example, we can imagine that the impacts of these early scientific constructions produced certain ways of thinking and feeling for those constructed as ugly and degenerate. Although, according to Katherine McKittrick, science socially produced the human, "it is also the epistemological grounds through which racial and sexual essentialism is registered and lived. These research foci and themes, for the most part, tend to underscore the long-standing prominence of scientific "facts" developed between the eighteenth and nineteenth" centuries (2015, 148). Similarly, Black American writer Ta-Nehisi Coates writes, for example that, "racism is a visceral experience, that it dislodges brains, blocks airways, rips muscle, extracts, organs, cracks bones, breaks teeth. ... You must always remember that the sociology, the history, the economics, the graphs, the charts, the regressions all land, with great violence, upon the body" (2015, 10). So, too sexism produces many violences that stick to and shrink the body. And yet violence is also felt affectively in the form of pain, agony, and yearnings, and these feelings are important to understand the full extent of a knowledge system that conceived and produces the human and the non-human, the person and the non-person.

Before the reading, the writing and arithmetic: Classifying and quantifying the human

In the Caribbean, I grew up hearing and playing jump rope, or what we refer to as skipping, to a version of "School Days" written in 1907 by Will Cobb and Gus Edwards. Part of the original song goes, "reading and 'riting and 'rithmetic, taught to the tune of the hickory stick." In the Caribbean, the words and rhythm of the song are rearranged to the more up-tempo beat of young girls singing as they jump to the hum of the skipping rope hitting the ground and then lagging in the air for a few seconds. Our version went,

> "Mr. Jones teaches the college the best he can (short pause)
> For reading and 'riting and spelling and 'rithmetic (short pause)
> He never forgot how to use his whip (short pause)"

The remainder of the rhyme was then followed by a recitation of the months of the year with short pauses had between each month -January (short pause), February (short pause), and so on. In our Caribbean childhood version of the song, hickory stick was replaced by whip, but the sentiment remains that learning should be enforced through the use of violence. That we, young girls from early childhood to late adolescence, would sing along and skip to songs that espoused corporal punishment as a way to enforce learning – a lash on the buttocks with a belt also known as "a strap," or a rap with the ruler on the knuckles – is important to understanding the disciplining power which has been built into our knowledge system over centuries.

In this section of the chapter, I embark on reading the academy from its origins. I demonstrate the education system's construction of and active complicity with the ugliness of artificial standards of classification dressed up with the status of so-called scientifically measured "proof" of white, male superiority in body and mind. Reading in the song mentioned above comes first. However, because the work I am engaging is subversive and seeks to counter the ways in which proof was had through violence, I seek to demonstrate the ways in which the Black and Brown and female body was first measured, and then violently written into history, turning the sequencing of "reading and 'riting and 'rithmetic" on its head through a good read which seeks to counter this violence.

In his *Two Treatises of Government*, philosopher John Locke prescribes a formulation of political society in which, according to da Silva, property ownership and the liberal subject as we know it today in the United States was configured through that construct "of the American continent as 'empty land ... as 'wild woods and uncultivated waste ..., left to nature, without any improvement ...' ... as if its inhabitants had failed to exceed the command of the law of nature (the law of reason) and act upon nature to produce more than that which is necessary for the perseveration of human life" (2007, 204). According to Locke, white men could lay claimed to this property by virtue of their labor. He writes

> The Labour of his Body, and the Work of his Hands, we may say, are properly his. Whatsoever then he removes out of the State that Nature hath provided, and left it in, he hath mixed his Labour with, and joined to it something that is his own, and thereby makes it his Property. It being by him removed from the common states Nature placed it in, it hath by this labour something annexed to it, that excludes the common right of other Men (1689).

Literally ascribing godlike power and prerogative to white men, Locke's basic principle and the idea of the autonomous, free will of the individual is the foundation of the notion of the social contract and modern liberalism, whereby if the individual together with others elects, they are free to form a state of limited powers to determine rules. However, as we know, "[p]olitical and social rights are distributed differently, not individually; they are based on constructions of the value of different groups to the nation" (Smith 2016, 32) such that for Locke, the labor of the enslaved, for example, did not grant them property ownership or accumulation and inclusion into the social contract. In fact, Charles W. Mills observes about what he refers to as the racial contract, the refusal to allow non-whites into the body polity that was created (Mills 1997), and Carole Pateman describes the analogous sexual contract (1988), which "constituted the female body as an unsuitable occupant of the body politics, [just as] certain racialised bodies were also deemed unsuitable participants of the politic" (Puwar 2004, 21).

Locke, also an investor in colonialism and the slave trade, most notably through England's Bahamas adventure, was able to amass great wealth (Perry 2018, 18), and owned according to the historical record, "at least £600 in Royal African stock" (Wilder 2013, 50). Not only was he regarded as one of the foremost influential enlightenment thinkers but he also (co)wrote the laws for the Carolina Charter of 1669 in which he stated, "Every freeman of Carolina shall have absolute power and authority over his negro slave, of what opinion or religion soever". Prior to Locke, Thomas Hobbes also theorized about the social contract based on the reasonableness of the individual to enter said contract, also justifying monarchial rule. A subject's reasonableness could be ascertained based on their education; and Hobbes, in *Leviathan* (1651), theorizes that educating the masses was the duty of the sovereign, as this education would play the role of helping to avoid civil war and maintain peace and order through instilling submissiveness in the masses. In addition, according to Teresa Bejan in her article "Teaching the *Leviathan*: Thomas Hobbes on education," Hobbes "distinguish[es] between different forms of teaching appropriate to different sections of the population" (2010, 617) where the elite classes learned in the university, and others received education through the church. In contrast to Hobbes, Locke in his 1693 treatise on education, *Some Thoughts Concerning Education*, written for the gentry class, proposed that the responsibility for education should be on parents. In this treatise Locke also writes, "Shame has in children the same place as modesty in women, which cannot be kept, and often transgressed against." Locke also writes of the native peoples in his essay on "Study," "perhaps without books we should be as ignorant as the Indians, whose minds are ill-clad as their bodies" (1667). Both Locke and Hobbes "wrote their theories of education in the tumultuous context of struggles around late feudalism and early capitalism [when] ... education emerged as part of primitive accumulation - that is, the creation of the preconditions for capitalist relations, which involved ... colonial dispossession and enslavement of Indigenous peoples, military suppression of peasant rebellions, and the degradation of women, seen most brutally with the execution of thousands of so-called witches" (Meyerhoff 2019, 151). These educational philosophies meant that women and racialized others were incorporated into universities as other than rational knowledge producers. According to Nirmal Puwar, women and non-whites represented "the negative side of the binaries of nature/culture, body/mind, affectivity/rationality, subjectivity/objectivity and particularity/universality. Conversely, because somatophobia is central to the definition of whiteness and maleness, both of these identities are defined as an absence of the bodily, a transcendence of the bodily into the realm of rationality, culture and enlightenment" (2004, 142). And as such, the non-white and non-male were designated as mere bodies to be used for the purposes of "rational" entities.

By this mode of reasoning, enslaved people then could not enter into the social contract as anything other than "exchangeable commodity[ies] in the eyes of traders, enslavers, and doctors" (Berry 2017, 4) and were likewise incorporated into the university as things to be measured, prodded, pried, dismembered and dug up. The thought was, according to Wilder, that "human beings occupied fixed racial categories with biologically determined fates [and] [a]cademics were well positioned to make this argument" (2013, 272). Educational and medical professions were thus given permission to examine the bodies of the enslaved and other designated non-humans which "became a Frankenstein's monster *avante le coup*: a collection of parts each one of which the surgeon could detach, reattach, delve into, and bring into view *as* parts. Knowledge became dependent upon the 'exposure of physical detail,' and the parts did not add up to an integrated whole (Crawford, 1996: 67)" (Chinn 2008, 18-19). These measurements extended into the wider society, as the university itself was positioned as a critical arm of the colonial state. Take, for example, that the enslaved were counted as 3/5th of a person in the US constitution – this accounting meant to benefit white landowners in that the more bodies they could count, the more power they could claim in the larger body politic – and according to Chief Justice Taney in the 1865 Dred Scott ruling, the Black man "had no rights which the white man was bound to respect." As I will discuss further on, the political designation of Black bodies as incomplete parts extended further into the realm of academic sciences.

With regard to women, it is important to note that the College of William and Mary incidentally opened around the same time as the Salem witch trials were happening. We can connect these two events in terms of their purpose and effects. William and Mary were expressly intended to civilize the "savages" and "heathens" (Wilder 2013, 156). According to bell hooks, the trials "were an extreme expression of patriarchal society's persecution of women" (1981, 31). As the story goes, these so-called witches were mostly white women (not all) who, while they were "mythologized as pure and virtuous" (hooks 1981, 31) as well as vulnerable and in need of protection, having no business in the world of men, money and talk of sex, happened not to conform to society's way of being. Unmarried and childless women in particular violated society's rules and mores on the basis of being female bodies which did not "produce" in the prescribed manner (married sex, legitimate children). The trials "were a message to all women that unless they remained within passive, subordinate roles they would be punished, even put to death" (hooks 1981, 30). As a Black Caribbean woman, I am connected and drawn to the history of the Salem Witch trials through one of its scapegoats Tituba, and the fictionalized story based on the scant historical records of her life and testimony at the trials told by Caribbean writer

Maryse Conde. As universities and colleges were being established on the American continent in the late 1600s, it was also established that women were not considered fit for the life of the mind. Females were "defined as representing all that the social contract in the political realm sought to exclude, that is, emotion, bodies, nature, particularity and affectivity" (Puwar 2004, 142). Pateman refers to what she calls the "[t]he disorder of women" as a condition used to justify their political disenfranchisement and their exclusion from the academy (1990, 4). The same biological and scientific methods used to justify the "taming" of the "savages and heathens" are the same that framed the moral disorder of women and was responsible for creating a system that excluded both women and Black and Brown people as outside of being capable of producing rational thought.

For white women, being excluded from the province of the university and characterized as irrational was also part of the colonial process understood as happening and "located within the context of ... the (hetero)normalization of the propertied and/or bourgeois-white family and the gendered/sexualized structures it fabricates and authorizes" (Santiago-Valles 2003, 219). Locke regarded the white woman as without personhood, solely having citizenship by virtue of her husband or father who "was the one possessed of full personhood" (Perry 2018, 23) empowered to protect, that is, regulate/punish those "citizens whose political representation was held by the family patriarch (white women and children)" (Smith 2016, 44). In other words, the same system that ordered the hierarchical dichotomies between the human and non-human was also responsible for legitimating white male superiority over white women. According to Breny Mendoza, as "[t]he human itself was bifurcated: as creatures closer to nature, emotional rather than rational, bound to the animal function of reproduction, European women were lower than men in the great Chain of Being, yet they were still human, marked by culture. Civilized gender involved a hierarchy that subordinated European white women to European white men, but still marked a gulf between colonizer and colonized" (2012, 18). Unlike racial subjection, according to da Silva, "gender subjection ... does not presuppose a scientific account of bodily difference" (2007, xxviii) as non-human. This is a distinction of note as while those racial/ethnic Others were constructed as non-human, white women were constructed as without *personhood*. Patriarchy as a mode of power circumscribed women as human to the domestic sphere, and also placed her body "under the scrutiny of scientific tools in the nineteenth century" as a non-person (2007, xxviii). French philosopher Michel Foucault has theorized that with modernity came the creation of "docile' bodies" which are under constant surveillance "described, judged, measured, compared with others ... trained or corrected, classified, normalized, excluded" (Chinn 2000, 4). So that basically white women were categorized under a state of conditional "citizenship" bestowed by intimate affiliation with specific white males and the laws governing those relationships, derived from particular philosophies.

For non-white women, classification was even more extreme. While the system relegated white women from the public, they were still seen as sensible and civilized enough for homemaking and childrearing: whereas non-white women were not even to be unsupervised in those activities and were often denied access to basic literacy. William Faulkner wrote of Black women, "But now and then a negro nursemaid with her white charges would loiter there and spell them [the letters on the sign] aloud with that vacuous idiocy of her idle and illiterate kind" (1932, 59). Black women have "consistently been written to inhabit the *public* ... place produced by scientific strategies where her body is immediately made available to transparent male desire but where her desire (passion, love, consent) is always already mediated by her double affectability. The result is that she is constructed as the subject of lust; hers is a dangerously unproductive will because it is guided by nothing but that which human beings possess as being ruled not even by the 'laws of [divine] nature,' the preservation of life" (da Silva 2007, 266). Black (and Brown) women, therefore, bore/bear both the marks of the non-human and the non-person such that they were and continue to be doubly affected by racists sexist ideology.

What the above demonstrates is that race and gender as socially con-structed categories work together to produce lived experiences under the conditions of modern capitalism (Santiago-Valles 2003, 222). Many deco-lonial scholars, including Latin American scholar Anibal Quijano, focus on the colonial conquest of the Americas by Europeans to argue that the coloniality of power is intimately linked to the creation of races, raciali-zation of labor, and racial dominance under a system of global capitalism arising under colonization and existing today, and that race and the division of labor are not only linked structurally but reinforce each other (2008). In her criticism of Quijano's lack of attention to gender, decolonial feminist María Lugones extends the idea of coloniality of power to consider gen-der as well as race, and their intersections and formations under colonial capitalism. Lugones builds upon Quijano's coloniality of power to develop what she calls the "colonial/modern gender system" to demonstrate how "'coloniality' does not just refer to 'racial' classification. It is an encompass-ing phenomenon since it is one of the axes of the system of power and as such it permeates all control of sexual access, collective authority, labor, subjectivity/inter-subjectivity and the production of knowledge from within these inter-subjective relations" (2016, 3). According to Lugones, this system of coloniality "understood race as gendered and gender as raced in par-ticularly differential ways for Europeans/'whites' and colonized/'non-white' peoples" (12). Double colonization was imposed in the case of non-white women who "were understood as animals in the deep sense of 'without gen-der,' sexually marked as female, but without the characteristics of feminin-ity. Women racialized as inferior were turned from animals into various modified versions of 'women' as it fit the processes of Eurocentered global capitalism" (13).

The development, centralization, and deployment of Eurocentric secularized knowledge was/is critical to colonial power (McLaren 2017, 22–24). This knowledge is responsible not only for dehumanizing and degendering Black and Brown men and women, but for reducing and controlling gendered white women in the private sphere to "reproduce the class, and the colonial, and racial standing of bourgeois, white men" (Lugones 2016, 15). As such, the gender hierarchy established during colonialism ensured that women's entire lives were lived in subordination to white men in public and private, including in the realm of knowledge and other means of production (Lugones 2016, 15). In this sense then, according to Lugones, the construction of Black and Brown people as (public) property and without gender and "the reduction of gender to the private, to control over sex and its resources and products is a matter of ideology, of the cognitive production of modernity" such that public knowledge production is gendered (Lugones 2016, 11–12), and raced; that is, not produced by Black or Brown people or white women. Building on Lugones' theorizing, decolonial feminists globally have been calling for the adoption of a decolonial approach to knowledge production, which is anti-sexist, anti-racist, and anti-capitalist (Mohany 2003), which pays attention to the effects of colonization on ways of knowing, and can expose relations of power implicit in knowledge creation (McLaren 2017).

I apply scrutiny and perform "reading" on the ways in which white colonial males have held out themselves as knowledge producers in the academy, justifying their supremacy through technologies of knowledge using (in the words of Ralph Cintron) "discourses of measurement" which are held out as impartial, universal standards of accuracy and modernity. Many examples come to mind. Wilder, who thoroughly documents the US university's connection to colonialism, slavery and genocide in his historiography of the university *Ebony and Ivy: Race, Slavery, and the Troubled History of America's Universities* writes, "The first five colleges in the British American colonies - Harvard (established 1636), William & Mary (1693), Yale (1701), Codrington (1745) in Barbados, and New Jersey (1746) - were instruments of Christian expansionism, weapons for the conquest of indigenous peoples, and major beneficiaries of the African slave trade and slavery" (2013, 17). In addition to these, other universities and colleges such as "the College of Philadelphia (University of Pennsylvania, 1749), King's (1754), the College of Rhode Island (Brown, 1764), Queens College (Rutgers, 1766) ... Even Dartmouth College (1769) in New Hampshire" relied "upon the generosity of the colonial elite," including merchants and slaveholders (Wilder 2013, 49). By the nineteenth century, public colleges in the United States, including many in the south, for example, state universities from South Carolina College in 1801 and the University of Virginia 1819, were founded by and received funding from men whose fortunes were amassed from racists mercantile ventures such as cotton and other commodity trades, which directly profited from slavery.

US academic institutions were (and are) "the intellectual and cultural playgrounds of the plantation and merchant elite" (Wilder 2013, 138). In large part, they did not include women or non-white people. These marginalized populations remained uneducated or attended segregated educational institutions like Mount Holyoke Female Seminary, founded in 1837, or Cheyney University, which opened that same year for Black people. From the establishment of Harvard University all the way through to the end of slavery, universities were funded by profits from the labor of enslaved people. Graduates of universities not only fought the native peoples and took their land but were instrumental in the trade of native peoples and enslaved Africans to the south and in the Caribbean; and some became plantation owners in the Southern US (Wilder 2013). And Northern whites were also implicated and benefitted directly from slave labor. The College of William and Mary in Virginia was opened by way of a charter granted by King William III and Queen Mary II and funded through "the profits of slave labor, assigning a duty of a penny per pound on tobacco exported from Virginia and Maryland to support a president and professors" (Wilder 2013, 42). Eli Meyerhoff quotes the 2001 "Yale, Slavery and Abolition" report which describes how John C. Calhoun, the seventh vice-president of the United States, "had been a student at Yale with his tuition paid by profits from enslaved people's labor, and went on to gain wealth and political power as a slave plantation owner, and became a statesman who wielded 'enormous political influence on the preservation of slavery'"(2019, 9). Meyerhoff observes that in "1930, Yale University decided to name 'Calhoun College' in his honor. Profits from slave labor provided much capital for Yale's first scholarships, early buildings, and endowment, and Yale's campus was itself a site of slave labor" (9).

Echoing Meyerhoff, Wilder writes about how King's college was a merchant's College enrolling approximately "ninety sons of the commercial class," that is, Atlantic traders, in its first two years – more than any other college (2013, 67). Wilder further observes that US universities and "[c]olleges were imperial instruments akin to armories and forts, a part of the colonial garrison with the specific responsibilities to train ministers and missionaries, convert indigenous peoples and soften cultural resistance, and extend European rule over foreign nations" (33). In fact, with the expansion of the African slave trade, higher education in the Mid-Atlantic and in New England realized a major expansion as well (Wilder 2013, 70), as the wealth generated by profits from colonial slavery paid for building campuses and shoring up college trusts and endowments. With that, and with slaveholders holding positions as college presidents, campuses had no choice but to conform to the "demands of slaveholding students as colleges aggressively cultivated a social environment attractive to the sons of wealthy families" (Wilder 2013, 77). The education, networks and, by extension, potential wealth generating opportunities and prestige provided to these wealthy white men through attending these institutions of higher learning "carried the American academy into modernity" (Wilder 2013, 111).

The 1862 Morrill Act, which made provisions for federal funds to create colleges to educate the laboring classes and set aside 30,000 acres of land confiscated from native peoples for each state to establish colleges that taught agriculture and mechanics to the laboring classes, funded the opening of many other schools (Meyerhoff 2019, 202). However, long after emancipation, there was strong opposition to developing these resources to go to Black people. In 1903, Senator Ben Tillman from South Carolina "warned that the colleges and schools for Black people in the South would lead inexorably to racial conflict. Designed to equip 'these people' who, in his eyes, were 'the nearest to the missing link with the monkey' to 'compete with their white neighbors,' these schools would ... 'create an antagonism between the poorer classes of our citizens and these people upon whose level they are in the labor market'" (Davis 1983, 124). I'm noting here something which continues to this day, how various forms of "class consciousness" can be used as a shield by white people to necessitate the further sub-classing of Black workers in the labor market. Dehumanizing regard of Black workers is explicit for those who cannot regard them as emancipated workers entitled to earn a living and advance socio-economically.

During Reconstruction, a few universities and colleges were integrated or solely dedicated to the education of Black and Brown people (and/or white women). Christi M. Smith gives three examples of integrated academic institutions in her book *Reparations & Reconciliation: The Rise and Fall of Integrated Higher Education* – Oberlin college founded in 1833, Howard University, whose first class in 1867 comprised of five white women, and Berea College, which in 1872 had an explicit policy that allowed interracial dating. However, opposition to inclusion continued, and this progressive movement toward diversity was short lived (Smith 2016, 6). Smith writes that post-emancipation it was thought by conservatives that "education could provide a means of "discipline" and white control over blacks" (48-49) while American reformists argued that education was "a means to produce a new class of voters and a site to encourage and organize competition between blacks and poor whites" (50). Clearly, these diverging views resulted in tension and Smith writes that essentially, "education gained support because elites viewed schools as a sorting device to distribute certain forms of citizenship rights to certain groups ... organized with the mandate that schools produce the right kinds of citizens for a new, stable and unified republic" (62). While the first Morrill Act provided aid for public universities, it failed to implement racial integration and, most institutions failed to substantively include women. In fact, W. E. B Du Bois, as quoted in Robinson, stated that the attack on reconstruction was *led* by the universities, with Columbia and Johns Hopkins at the forefront (Robinson 2000, 188). The second Morrill Act in 1890 sought to rectify this underpinning of inequality, but rather than forcing public universities to develop integration programs, a separate tier of federal funding was awarded to private historically black colleges and universities (Smith 2016, 241).

The history of the colonial university is one of exclusion of non-white people and women while using the bodies and labor of enslaved people to build their campuses and their academic disciplines. According to Wilder, American university campuses stand as monuments of slavery, as enslaved Black people not only built "Thomas Jefferson's intellectual monument: the University of Virginia" (2013, 137) but racist curiosity about their biology and genetic make-up also provided fodder for white students' intellectual development, study, and development of disciplines. Wilder writes

> Students ... crafted a science that justified expansionism and slavery - a science that generated broad claims to expertise over colored people and thrived upon unlimited access to nonwhite bodies. They ... redefined truth ... deploy[ing] science to prove the prophecies of the Bible, and now, with similar vigor, they pursued the visible and manifest truths of the material world. Race did not come *from* science and theology; it came *to* science and theology. ... Science and theology deferred to race, twisting and warping under the weight of an increasingly popular and sweeping understanding of human affairs that tied the social fates of different populations to perceived natural capacities (182).

With the end of reconstruction and the ushering in of the Nadir, Sarah E. Chinn observes that white supremacy conjured up "a hard-and-fast methodology to identify black bodies, count them, and fix them in place. ... reading the body as evidence" (2000, 4). Quoting Davis, Chinn recalls that in 1890, reminiscent of the 3/5th clause and no doubt of the "one drop rule," that the US Census required enumerators to "record the *exact* proportion of 'African blood'" in the counted, "relying on visibility" (4). Of course, as Chinn points out, the census is already an instrument of disembodied quantitative measurement, a single document which collates bodies as material evidence for the enactment of policies around schooling, welfare and immigration, the allotment of funding for programs and initiatives, and the marking of electoral districts. She goes on to write that, "Not only were they to count how many black people there were; they also had to quantify *how* black they were in measurable fractions" (4). The academy led this effort during this period, and as the place of science became increasingly tied to the institution, and defined by specific fields and disciplines, "scientists represented themselves as searchers after the truth" and increasingly came to be held in high regard in American culture (Chinn 2000, 17) as the ones who could find out the truth about human life.

Some examples of how the academy made and remade the body in order to substantiate exclusion through the use of so-called scientific data "to count, measure, quantify, record" (Chinn 2000, 4) can be seen when we examine the period of the nineteenth century to the late twentieth century. Take, for example, the bestselling text *The Bell Curve*, published by Charles Murray in 1994 and which makes the claim that social differences stem from the higher fertility rates of genetically less intelligent and racialized groups, where "Black mothers have been thought to pass down to their offspring the traits that marked them as inferior to any white person" (Roberts 2017, 8). Murray's white supremacist thesis comes from a long line of racist scientific measurements of bodies which have pervaded the academy since the 1800s, particularly by eugenicists scientists. In her article entitled "Mama's Baby, Papa's Maybe: An American Grammar Book," Black feminist Hortense Spillers, making the case that the enslaved were what she calls "a living laboratory," reproduces for us an advertisement for what she calls "[a]ssortments of diseased, damaged, and disabled Negroes, deemed incurable and otherwise worthless ... *bought up*, it seems ... by medical institutions, to be experimented and operated upon, for purposes of 'medical education' and the interest of medical science" (1987, 68). According to Diana Ramey Berry, universities also dug up the bodies of Black people to fuel the cadaver trade. In describing several examples, she writes, "From 1848 to 1852, the Medical College of Georgia used 'resurrection slaves...' to rob graves from local cemeteries ... and then reinternment after the medical students and faculty were finished with the dissection" (2017, 169). The bodies of enslaved people could not even escape the brutal measurements and dissection of academic science, even in death.

Dorothy Roberts, in her book *Killing the Black Body: Race Reproduction and the Meaning of Liberty*, does a fantastic job of demonstrating the link between racist science and the propagation of eugenics programs. She develops an indictment of Francis Galton, who popularized the "argument for affirmative state intervention in the evolutionary process" (2017, 59) and coined the term eugenics. Roberts' work catalogs the various ways that Black women's bodies were made and remade by science over centuries, as a means of refining the precision of racist, sexist theorizing that used their bodies as evidence and for experimentation, and which undergird many of our laws and policies today. Eugenicists like Galton used biology, statistics, philosophy, neurology, mixed in with religion, mere conjecture based on stereotypes, and proselytizing, to develop their racialized and gendered science. According to Chinn, "Galton's successful crusade for quantification changed the way biologists, botanists, astronomers, medical researchers, and a host of other scientist approached the collection and demonstration of evidence" (2000, 18). In his 1869 text *Hereditary Genius,* Galton claimed that he could statistically measure intelligence. He deployed a number of mathematical calculations and, in his 1889 book *Natural Inheritance* wrote that "[s]ome people hate the very name of statistics, but I find them full of beauty and interests" (Kevles 1998, 17). Galton used his measurements to make claims about bodies, believing "intelligence and character were transmitted by descent" (Roberts 2017, 59-60) and that defective genes could be found in Black people, immigrants, those marginalized by gender and sexuality, the poor, and those with disabilities; and he sought to prove this through statistics. He, according to Chinn, "believed that what stood in the way of knowledge was less the lack of data than the way those data were organized. Statistical analysis would render more precision in all kinds of scientific endeavors, illuminating and amplifying human knowledge of natural and mathematical processes" (2000, 17). As Stephen Jay Gould states, "[q]uantification was Galton's god" (1996, 108) and Galton ascribed powers to it as if they were bestowed by God upon white men.

By the late 1800s, eugenicists were using data analysis schemes for their purpose. For example, Charles Warren, Robert DeCourcy Ward and Prescott F. Hall from Harvard, who founded the Immigration Restriction League in 1894, used "scientific," statistical or medical research and methodologies to prove that immigrants from southern and eastern Europe were racially inferior to Anglo-Saxons, threatened the American way of life and would bring with them poverty and organized crime at a time of high unemployment (Strings 2019, 147). In other words, these bodies posed an economic threat. Iris Marion Young describes an *episteme* by which "bodies are both naturalized, that is, conceived as subject to deterministic scientific laws, and normalized, that is, subject to evaluation in relation to a teleological hierarchy of the good. The naturalizing theories were biological or physiological, and explicitly associated with aesthetic standards of beautiful bodies and moral standards of upstanding character" (1990, 127). Aesthetic standards based on white, male bodies were used to devalue non-white people and women as their bodies were reduced to statistical models to dis/prove normality, "and rendered visible only as a collection of physical features (skin color, hair texture, skull size, and so on)" (Chinn 2000, 18). By the end of the nineteenth century, for example, experts in multiple scientific areas of expertise including, medical doctors and public health experts, biologists, psychologists, sociologists and anthropologists, were all able to produce some type of "scientific" evidence about the Black woman's embodiment and sexuality (Hammonds 2012, 172).

According to Megan Glick, "While the connection between intelligence and humanity was well established by earlier versions of racial science, the early twentieth century saw the formalization, standardization, and bureaucratization of intelligence as a determinant in human hierarchy" (2018, 62). According to Kevles (1998, 17), by the end of the nineteenth century there was a shift in the natural sciences from "mere data gathering" to mathematical qualification as a result of Galton's contribution. The "allure of numbers" (Gould 2996, 106) was such that following Galton's work, wealthy financiers backed the work of eugenic scientists such as Charles Davenport, a biologist who attended and taught at Harvard and curated the Museum of

Zoology at the University of Chicago. Davenport too theorized about the heredity of so-called defective genes and the importance of race to determining behavioral traits. Roberts writes of Davenport: his

> Cold Spring Harbor project supplied the burgeoning American eugenics movement with adherents and research: it trained and dispersed over 250 field workers, published the *Eugenical News*, and disseminated bulletins and books about the reduction of hereditary degeneracy. As Davenport conducted scientific research, eugenics became the vogue across the country. Ordinary Americans attended lectures and read articles in popular magazines on the subject. Those devoted to studying eugenics joined organizations such as the American Eugenics Society, the American Genetics Association, and the Human Betterment Association. *The Reader's Guide to Periodical Literature* listed 22 articles under "eugenics" between 1910 and 1915, making it one of the most referenced topics in the index. At most American colleges courses on eugenics were well-attended by students eager to learn how to apply biology to human affairs. The American Eugenics Society reached a less erudite audience by sponsoring Better Babies and Fitter Families contests at state fairs across the country (2017, 62).

The popularization of eugenic science and the dissemination of its standards of "fitness" among the educated and uneducated classes as seem through the various eugenics societies, worked not only to proliferate these standards but to make measuring up to these standards aspirational regardless of class or economic status. These standards could then be applied in greater society to shore up resistance to Anglo-Saxon solidarity with Others.

It is important to note how during this time, according to da Silva, "the concept of the cultural would inform a scientific project for which man and his social configurations constitute the privileged object ... [and] operate as a signifier of globality, that is, as a strategy of engulfment, a political-symbolic strategy that hides and announces spatiality by writing the actual and possible relationships between the coexisting human collectivities it maps as governed by the productive *nomos*" (2007, 131-132). It is through these academically and "scientifically" manufactured knowledges that racial and sexual difference came to also stand in for social and cultural markers of difference globally (da Silva 2007, 133), used by the United States in the late nineteenth century and the early twentieth century to justify imperialism based on discourses of "civilization" (Glick 2018, 32) that imbued racialized others including Black people and immigrants as mentally underdeveloped. The academy has argued for and provided the "evidence" necessary to justify the United States's increased presence in the global colonial project, amassing an empire through a rhetoric of humanitarian concern, charity, and humanizing, and civilizing nations as a matter of duty including through military intervention.

At the same time eugenics was growing in popularity, there was also another academic theory in the form of the IQ test, which theorists claimed "could quantify innate intellectual ability in a single measurement" (Roberts 2017, 63). According to Roberts, at the beginning of the twentieth century, physical measurements of the brain through techniques such as craniology were replaced by "mental tests" as a so-called objective, unbiased and quantifiable way of ranking ability through genetic transfer and thus determining human inferiority and superiority, decrying race mixing. Many psychologists from schools like Harvard and Princeton used these IQ tests developed by French psychologist Alfred Binet[1] as a diagnostic tool, to show how Black and Brown people and immigrants from Southern and Eastern Europe were the intellectual subordinates of Anglo-Saxon whites (Roberts 2017, 63). For example, "psychologists H.H. Goddard, L.M. Terman, and R.M. Yerkes transformed the IQ into *the* measure of a race-, ethnicity- and class-specific, hereditary, reified, unitary attribute called intelligence. Individual scores could locate a person along a fixed numerical scale that ranged from 'genius' (above 145 points) to 'average' (100 points) to 'high-grade defective' (that is, fit only for unskilled work: 75 points); the scale ended at 'feeble-minded.'" (Chinn 2000, 52). The works of these and other scientists were given legitimacy through publication in academic journals and taken up by government officials who used them in making political decisions about migration, marriage, and social welfare programs. A commentary in the *Yale Law Review* on the 1923 new edition of the book *The Passing of the Great Race* by Madison Grant anthropologist at the American Museum of Natural History, written by E. Huntington, summarizes this well when the reviewer observes, "America is seriously endangering her future by making fetishes of equality, democracy, and universal education."

These measurements and the normalizing scientific gaze continued to have a damaging impact on non-white people and women all the way up until today in the form of, for example, sterilization and forced institutionalization. Scientific racism in the form of mental and physical measurements predisposed white Americans to believe theories of biologically determined inherited character and behavior either as normal or deviant, and as such to condone violent medical interventionist practices like sterilization for the overall well-being of the society (Roberts 2017, 61). Not only sterilization but the case was made by many including Harvard geneticist Edward East, that providing "prenatal care and obstetric services to the poor through clinics and public hospitals was 'unsound biologically' because it 'nullifie[d] natural elimination of the unfit.'" (Roberts 2017, 65). According to Chinn "dissection and detail were key to the processes of creating evidence of identity, and we can see these features in some of the crucial developments of the contemporary era" (2000, 19). What we learn from the above is that bodily abstraction for measurement and quantification through hegemonic knowledge production "renders it [the body] legible as a sign of something else, not itself: patterns, qualities, trends, predictable processes" (Chinn 2000, 5). This process, in turn, sets up normalizing frames through which we look at society and those within it either as normal or not, making it almost impossible to see, think, and act outside of these frames (Mbembe 2015, 10). There is a through line that spans the scientific justifications of slavery directly to the science used to propagate eugenics to sustain a societal order important to the capitalist structure. For Sandra Harding, knowledge development and the trajectory of disciplinary paradigms "are fully part of their historical era, bearing the fingerprints of these eras and the subsequent ones that practice and maintain them in their cognitive core. Prejudices have no doubt bled through data which then circled back to the same prejudices—an unbeatable system that gained authority because it seemed to arise from meticulous measurement" (2001, 292).

This deliberately constructed system created what Glick refers to as "the omnipresence and slipperiness of eugenic ideology" (2018, 13). Roberts makes this clear as she points out that the eugenics movement found major support and backing from "the nation's wealthiest capitalists, including the Carnegie, Harriman, and Kellogg dynasties" (2017, 61) showing a direct correlation between the role of the university to upholding the capitalist social order. The through line is even more pronounced as, according to Roberts, "[i]t bears remembering that in our parents' lifetime states across the country forcibly sterilized thousands of citizens thought to be genetically inferior" (2017, 59). The impact of these theories continues in present day policies where instead of providing prenatal care and other health services for impoverished Black and Brown women, governments still choose to implement a policy of sterilization to access certain social services. This through line demonstrates the impact of how scientific measurement has consistently been used and promoted by the university to justify white patriarchal supremacy in public policy, to impact "the way Americans value each other and thought about social problems" (Roberts 2017, 103). Smith argues "that between the rise of nationhood in the nineteenth century, we see the rise of new foundation for citizenship – what one might call *jus laurea* – or civic and social inclusion based on education accomplishment. ... The struggle over education shaped ideas about the cultural deservedness of different groups, and these played into political adjudication of rights and opportunities based on group membership" (2016, 23–24).

Many of these theories developed in and perpetuated by the academy have since been exposed as false. And yet, to read the university is to see their footprints still existing within the modern academy, as Anglo-Saxon European normality and rationality are still presumed, and the U.S.

university's place as the home of the mind and intellectual and scholarly invention, accepted as fact (Mills 1997, 33–34). Today, "[e]ducational organizations generate categories, allocate status, and influence group boundaries … and the logics of these boundaries and of status and 'merit' are deeply contested and activated to serve particular organizational interests" (Smith 2016, 236). To be placed outside of the construct of what it means to be human and to possess personhood was not only to be constructed by others but to have the spaces of knowledge construction made unavailable to you. It resulted in racialized and gendered seemingly out-of-place-ness in a space that not only produces but validates, that is, normalizes these systems of production, knowledge construction and deconstruction and reinforces and reproduces a "narratively condemned status" as not human or rational and therefore not normal (Wynter 1994, 70). A whole set of "subjugated knowledges," that is "a whole series of knowledges that have been disqualified as non-conceptual knowledges, as insufficiently elaborated knowledges: naïve knowledges, hierarchically inferior knowledges, knowledges that are below the required level of eradication and scientificity" have been erased by the colonial nature of the academy (Foucault 2003, 7). Quoting da Silva, it is clear that

> [f]rom the initial deployment of racial difference as social scientific signifier, it has consistently rewritten post-Enlightment European social configurations and social subjects in transparency. On the one hand, it constructs the heirs of yesterday's natives as modern Calibans, "strangers" whose racial difference produces the affectable (unbecoming/pathological) moral configurations bringing about their subjection. On the other hand, it entails signifying strategies that engulf the globe – namely, "civilization," "modernization," and "globalization" – which retain as a presupposition the science of man's writing of Africa, Asia, and Latin America as subaltern global regions (2007, 181).

To recognize the overwhelming ways in which the social construction of race and gender has been produced and produces academic disciplines is to realize how any work by the academy to improve itself must engage critically with its colonial foundations. As decolonial scholars who currently work within these institutions and recognizing that "decolonization is always an unfinished project" (Memdoza 2012, 14) we also have to critically engage with this past if we are to envision a more just system, one which recognizes other, traditionally, globally produced knowledge.

As demonstrated above and throughout this chapter, bodies have historically been manipulated through the disciplinary mechanisms of the knowledge production process to classify them as normal or else non-human. This is a process itself which is a manifestation of social power since only certain bodies are classified as worthy to produce knowledge

(Foucault 1977). Decolonial scholar Achille Mbembe writes that the Eurocentric canon

> attributes truth only to the Western way of knowledge production. It is a canon that disregards other epistemic traditions. It is a canon that tries to portray colonialism as a normal form of social relations between human beings rather than a system of exploitation and oppression. Furthermore, Western epistemic traditions are traditions that claim detachment of the known from the knower. They rest on a division between mind and world, or between reason and nature as an ontological a priori. They are traditions in which the knowing subject is enclosed in itself and peeks out at a world of objects and produces supposedly objective knowledge of those objects. The knowing subject is thus able, we are told, to know the world without being part of that world and he or she is by all accounts able to produce knowledge that is supposed to be universal and independent of context (2005, 9).

Furthermore, the power of this prevailing system of expertise, not only regulates the subject but also results in self-regulating behavior through normalizing practices: these define the normal in advance, and then proceed to isolate, then subjugate and punish, anomalies resulting from that definition (Bartky 1990). These "normalizing effects" in turn shape how generations of individuals understand themselves, and their bodies and therefore measuring practices like the posture portraits described later in this text and the corresponding system of archiving anthropometric measurements and IQ testing, are best analyzed through an exploration of how they contributed to the development of bodies decades after these measurements were taken, including current classroom and test performance and their impacts. How do we understand this process as bound up in the current pressures to radically transform and include "bad bodies" (Bartky 1990), to produce by contrast "good," read normal, bodies who are worthy and can be what the society needs them to be and to take on the roles a hierarchal, capitalist society requires them to? What are the outcomes of this attempt at transformation and inclusion, including affective outcomes?

Analyzing these prejudices through a decolonial Black feminist lens helps us to consider the various marginalized identity positions occupied by those, who through colonial knowledge production used to justify their oppression and exclusion, are constructed as different. In Chapter 3 of this text, I examine university inclusion initiatives, with a specific focus on twenty-first century diversity initiatives, and thus I will show how the effects of colonialism on the normalization of hierarchal difference, and how what I define in the next chapter as the *politics of indifference* works in the academy to perpetuate the exclusion of gendered, raced and Othered bodies through measurement and data collection, while superficially espousing principles of inclusion and equality for the sake of appearing to be up with the times, and the affective impact of covert and overt oppression.

Note

1. The Stanford-Binet tests (although revised, most recent version in 2003) are still being taught to clinicians and educators and administered to children and adults.

References

Alcoff, Linda Martín. "Decolonizing Feminist Philosophy." In *Decolonizing Feminism Transnational Feminism and Globalization*, edited by Margaret A. McLaren, 49–75. London: Rowman and Littlefield International, 2017.

Alexander, M. Jacqui. *Pedagogies of Crossing: Meditation on Feminism, Sexual Politics, Memory, and the Sacred.* Durham: Duke University Press, 2005.

Alexander-Floyd, Nikol G. "Critical Race Black Feminism: A 'Jurisprudence of Resistance' and the Transformation of the Academy." *Signs: Journal of Women in Culture and Society 2010* Vol. 35, no. 4 (2010): 810–820. doi: 10.1086/651036.

Bartky, Sandra Lee. *Femininity and Domination: Studies in the Phenomenology of Oppression.* New York: Routledge, 1990.

Bejan, Teresa M. "Teaching the 'Leviathan': Thomas Hobbes on Education." *Oxford Review of Education* Vol. 36, no. 5 (Political and Philosophical Perspectives on Education Part 1 (October 2010)): 607–626. doi: .org/10.1080/03054985.2010.514438.

Berlant, Lauren. *Cruel Optimism.* Durham: Duke University Press, 2011.

Berry, Daina R. *The Price for Their Pound of Flesh: The Value of the Enslaved, from Womb to Grave, in the Building of a Nation.* Boston: Beacon Press, 2017.

Coates, Ta-Nehisi. *Between the World and Me.* New York: Spiegel & Grau, 2015.

Césaire, Aimé. *Discourse on Colonialism.* Translated by Joan Pinkham. New York: Monthly Review Press, 1972.

Chinn, Sarah E. *Technology and the Logic of American Racism: A Cultural History of the Body as Evidence.* London: Continuum, 2000.

da Silva, Denise Ferreira. *Toward A Global Idea of Race.* Minneapolis: University of Minnesota Press, 2007.

da Silva, Denise Ferreira. "Before *Man*: Sylvia Wynter's Rewriting of the Modern Episteme." In *Sylvia Wynter: On Being Human as Praxis*, edited by Katherine McKittrick, 90–105. Durham: Duke University Press, 2015.

da Silva, Denise Ferreira. "1 (Life)÷ 0 (blackness) = $\infty - \infty$ or ∞ /∞: On Matter Beyond the Equation of Value." *e-flux Journal* #79 (2017): 1–11.

Davies, Carole Boyce, Meredith Gadsby, Charles Peterson and Henrietta Williams. *Decolonizing the Academy: African Diaspora Studies.* Trenton: Africa World Press Inc., 2003.

Davis, Angela Y. *Women, Race, and Class.* New York: Vintage Books, 1983.

Davis, Angela. *Freedom Is a Constant Struggle: Ferguson, Palestine, and the Foundations of a Movement.* Chicago: Haymarket Books, 2016.

Doyle, Laura. *Bordering on the Body: The Racial Matrix of Modern Fiction and Culture.* Oxford: Oxford University Press, 1994.

Elliott-Cooper, Adam. "'Free, Decolonized Education' – A Lesson from the South African Student Struggle." In *Dismantling Race in Higher Education: Racism, Whiteness and Decolonizing the Academy*, edited by Jason Arday and Heidi Safia Mirza, 289–296. London: Palgrave McMillian, 2018.

Faulkner, William. *Light in August*. New York: Smith & Haas, 1932.

Foucault, Michel. *Discipline and Punish*. France: Gallimard, 1977.

Foucault, Michel. *Society Must Be Defended: Lectures at the Collège De France, 1975–76*, edited by Mauro Bertani, Alessandro Fontana, and François Ewald. New York: Picador, 2003.

Glick, Megan H. *Infrahumanisms: Science, Culture, and the Making of Modern Non/Personhood*. Durham: Duke University Press, 2018.

Goldie, Mark. *John Locke: Selected Correspondence*. Oxford: Oxford University Press, 2000.

Gould, Stephen Jay. *The Mismeasurement of Man*. New York: W.W. Norton & Company, 1996.

Hammonds, Evelynn M. "Toward a Genealogy of Black Female Sexuality: The Problematic of Silence." In *Feminist Genealogies, Colonial Legacies, Democratic Futures*, edited by M. Jacqui Alexander and Chandra Talpade Mohanty, 170–182. New York: Routledge, 1997.

Hanchard, Michael. "Identity, Meaning and the African-American." *Social Text, No. 24* Vol. 8, no. #2 (1990): 31–42. doi: 10.2307/827825.

Haraway, Donna. *Primate Visions: Gender, Race, and Nature in the World of Modern Science*. New York: Routledge, 1989.

Harding, Sandra. "After Absolute Neutrality: Expanding Science." In *Feminist Science Studies: A New Generation*, edited by Maralee Mayberry, Banu Subramaniam and Lisa H Weasel, 291–320. London: Routledge, 2001.

Harris, Cheryl I. "Whiteness and Property." *Harvard Law Review* Vol. 106, no. 8 (1993): 1707–1791.

Hobbes, Thomas. *Leviathan*. England, 1651.

Hong, Grace Kyungwon. "The Future of Our Worlds": Black Feminism and the Politics of Knowledge in the University Under Globalization." *Meridians* Vol. 8, no. 2 (2008): 95–115.

hooks, bell. *Ain't I a Woman: Black Women and Feminism.* Boston: South End Press, 1981.

Kevles, Daniel J. *In the Name of Eugenics: Genetics and the Uses of Human Heredity*. Cambridge: Harvard University Press, 1998.

Lethabo, Tiffany. "New World Grammars: The 'Unthought' Black Discourses of Conquest." *Theory & Event* Vol 19, no. 4 (2016) muse.jhu.edu/article/633275.

Locke, John. *Two Treaties of Government*. England: Awnsham Churchill, 1689.

Locke, John. *Some Thoughts Concerning Education*. England: University Press, (1693) 1902.

Lowe, Lisa. *The Intimacies of Four Continents*. Durham, NC: Duke University Press, 2015.

Lugones, María. "The Coloniality of Gender." In *The Palgrave Handbook of Gender and Development*, edited by Wendy Harcourt, 13–33. London: Palgrave Macmillan, 2016.

Maldonado-Torres, Nelson. "On the Coloniality of Being: Contributions to the Development of a Concept." *Cultural Studies* Vol 21, no. 2 (March 2007): 240–270.

Mbembe, Achille. *Decolonizing Knowledge and the Question of the Archive.* Africa is a Country, 2015. https://africaisacountry.atavist.com/decolonizing-knowledge-and-the-question-of-the-archive.

McKittrick, Katherine. "Yours in the Intellectual Struggle: Sylvia Wynter and the Realization of the Living." In *Sylvia Wynter: On Being Human as Praxis*, edited by Katherine McKittrick, 1–8. Durham: Duke University Press, 2015a.

McKittrick, Katherine. "Axis, Bold as Love: On Sylvia Wynter, Jimi Hendrix, and the Promise of Science." In *Sylvia Wynter: On Being Human as Praxis*, edited by Katherine McKittrick, 142–163. Durham: Duke University Press, 2015b.

McLaren, Margaret A. "Introduction: Decolonizing Feminism." In *Decolonizing Feminism Transnational Feminism and Globalization*, edited by Margaret A. McLaren, 21–61. London: Rowan and Littlefield International, 2017.

Mendoza, Breny. "Coloniality of Gender and Power: From Postcoloniality to Decoloniality." In *The Oxford Handbook of Feminist Theory*, edited by Lisa Disch and Mary Hawkesworth, 2–26, 112. Oxford: Oxford University Press, 2012.

Meyerhoff, Eli. *Beyond Education: Radical Studying for Another World.* Minneapolis: University of Minnesota, 2019.

Mills, Charles W. *The Racial Contract.* Ithaca: Cornell University Press, 1997.

Mohanty, Chandra Talpade. *Feminism Without Borders: Decolonizing Theory, Practicing Solidarity.* Durham: Duke University Press, 2003.

Pateman, Carole. *The Sexual Contract.* Oxford: Polity Press, 1988.

Pateman, Carole. *The Disorder of Women: Democracy, Feminism, and Political Theory.* Stanford: Stanford University Press, 1989.

Perry, Imani. *Vexy Thing: On Gender and Liberation.* Durham: Duke University Press, 2018.

Puwar, Nirmal. *Space Invaders: Race, Gender and Bodies Out of Place.* Oxford: Berg, 2004.

Quijano, Aníbal. "Coloniality of Power, Eurocentrism and Latin America," *Coloniality at Large* (2008): doi: 10.1515/9780822388883-009.

Roberts, Dorothy. *Killing the Black Body: Race, Reproduction and the Meaning of Liberty.* New York: Vintage Books, 2017.

Robinson, Cedric J. *Black Marxism: The Making of the Black Radical Tradition.* Chapel Hill: The University of North Carolina Press, 2000.

Santiago-Valles, Kelvin. "Some Notes on 'Race,' Coloniality, and the Question of History Among Puerto Ricans." In *Decolonizing the Academy: African Diaspora Studies*, edited by Carole Boyce Davies, Meredith Gadsby, Charles Peterson and Henrietta Williams, 217–234. Trenton: Africa World Press Inc., 2003.

Smith, Christi M. *Reparations & Reconciliation: The Rise and Fall of Integrated Higher Education.* Chapel Hill: University of North Carolina, 2016.

Spillers, Hortense J. "Mama's Baby, Papa's Maybe: An American Grammar Book." *Diacritics* Vol. 17, no. 2 (1987): 65–81.

Strings, Sabrina. *Fearing the Black Body: The Racial Origins of Fat Phobia.* New York: New York University Press, 2019.

Vaccaro, Annemarie and Melissa J. Cambra-Kelsay. *Centering Women of Color in Academic Counterspaces: A Critical Race Analysis of Teaching, Learning, and Classroom Dynamics.* London: Lexington Books, 2016.

Walcott, Rinaldo. "Beyond the 'Nation Thing': Black Studies, Cultural Studies, and Diaspora Discourse (or the Post-Black Studies Moment)." In *Decolonizing the Academy: African Diaspora Studies*, edited by Carole Boyce Davies, Meredith Gadsby, Charles Peterson and Henrietta Williams, 107–124. Trenton: Africa World Press Inc., 2003.

Wilder, Craig Steven. *Ebony and Ivy: Race, Slavery, and the Troubled History of America's Universities*. New York: Bloomsbury Press, 2013.

Wynter, Sylvia. "No Humans Involved: An Open Letter to My Colleagues." *Forum N.H.I Knowledge for the 21st Century: Knowledge on Trial* Vol. 1, no. 1 (1994): 42–73.

Wynter, Sylvia. "Unsettling the Coloniality of Being/Power/Truth/Freedom: Towards the Human, After Man, Its Overrepresentation – An Argument." *CR: The New Centennial Review* Vol. 3, no. 3 (2003): 257–337. doi: 10.1353/ncr.2004.0015.

Young, Iris Marion. *Justice and the Politics of Difference*. Princeton: Princeton University Press, 1990.

2 Yearning, the politics of indifference and the apathetic methodology of power

> Too often, it seems, the point is to promote the *appearance* of difference within intellectual discourse, a 'celebration' that fails to ask who is sponsoring the party and who is extending the invitations
>
> – bell hooks 1990, 54

> Our bodies are densely personal, written over with signs of the lives that inhabit them. It has been one of the triumphs of late twentieth-century feminist and queer activism and scholarship to insist upon the intimate interconnections of body and culture, body and intellect, body, and (dare I say it) soul.
>
> – Sarah Chinn 2000, 22

In the previous chapter, I examined how the scientific propagation of racist and sexist beliefs dressed up as empirical value-neutral science, determined that the modern subject would be, according to Denise Ferreira da Silva, constructed "in exteriority/affectability" (2018, 17). Scientists, theorists, philosophers, academic men of reason and rationality, developed and deployed scientific methods and a positivist approach to prove *a priori* held beliefs that anchored women to the affective and that determined non-white people to be lacking in humanity. Black feminist scholar Tamura Lomax stated it well: "the insignia of scientific validation corroborated by systematic study, observation, experimentation, and findings, endorsed and helped institutionalize a pervading and flourishing ideological structure. Science reconfigured racial and sexual re/presentation and fantasy as truth" (2018, 17). In this chapter, I argue that operationalizing these scientific systems was and is based on enormous amounts of indifference about the impact of scientific truth claims upon those whose bodies were/are measured and quantified. While this might seem like an obvious statement given the callousness of a racist patriarchal system, which produced the concept of clinical detachment as crucial to scientific study, it is worthwhile pointing out the importance of indifference as historically salient to the making of the scientific and academic community, which used racism and sexism as a tool

DOI: 10.4324/9781003019442-3

of exclusion. Pointing this out is especially significant at the current time where the university is said to be interested in acknowledging its past exclusions, is paying attention to inclusion, and in so doing, need to produce evidence of its commitment to said inclusion.

In our contemporary neocolonial society, even under those enterprises "characterized as 'development' policies dictated by U.S.-controlled financial institutions such as the World Bank and IMF, the concomitant displacement of laboring populations, and the transnational flow of goods, capital, and labor that is attendant to this condition" (Hong 2008, 102), the university continues to perpetuate significant violence against those it has historically excluded in order to continue benefiting from its role in the current iteration of the global political economy. The neoliberal, postcolonial state has emerged through the proliferation of field-based sciences and study based in the academy using so-called rational measurements to delimit the boundaries of the academy and to demarcate who belongs and who does not through the production of data. Some such data are generated through diversity and inclusion initiatives. These initiatives became popular after the profound changes which occurred in the university due to social movements of the 1950s, 1960s and 1970s, and are part of the resulting "retrenchment in reaction to these movements" (Hong 2008, 102) and after the landmark *Brown vs Board of Education* case of 1954. Many look to *Brown* as having resolved the issues of segregation and racism in the United States particularly with regard to school access and educational inclusion, and American children and youth are taught in schools that *Brown* was the end of the segregation chapter in the U.S. history. But as Black Feminist legal scholar Cheryl L. Harris makes clear, *Brown* has a "mixed legacy" in recognizing the unconstitutionality of legalized racial separation. Moreover, it also fails to address government responsibility and provide resources to remedy *de facto* separation. According to Harris, after *Brown*, the continued and consistent perpetuation of institutional privilege remained, but under the guise of pursuit of equality (1993, 1757).

During the 1960s and 1970s, social movements, including the racial liberation and women's movements, ushered new area studies programs such as women's studies and Black studies into the academy, largely due to student activism. According to Hong: "These student movements approached the University as an institution already implicated in a worldwide system of neocolonial and racialized capitalist exploitation. As such, their efforts were to change the very function of the University ... [as they] imagined ... a means of redistributing resources, producing counter-knowledges, and critiquing white supremacy and imperialism" (2008, 100). What these students found instead was the university was a formidable site for capitalist reproduction, and that instead of change, these academic institutions continued to perpetuate what Mendoza calls "diverse technologies of intellectual genocide" (2016, 15) and what Christina Sharpe describes as a "death-dealing episteme" (2014). These students found that, as Courney Moffett-Bateau explains, "the power of American ideology is so strong that it absorbs critique into its very definition of self. In this way ...celebrat[ing] the plurality of diversity without taking on its burdens" (2018, 87). As the saying goes, "the more things change the more they remain the same," and the university is the embodiment of this. The decision to respond by integrating student demands at the time, were met begrudgingly and with indifference, such that integration was less about obliging student demands for equity and justice and more about the institution's survival. The incorporation of different bodies and knowledge systems within the university spurred by student activism in the late 1900s did not change either the "conceptual, material, [or] affective architectures that reproduce the modern subject ... [instead the University] shift[ed] with new formations of the nation-state and global capital, as well as the dynamics that ar[o]se in efforts to resist their violence, and efforts in turn to co-opt and contain ... resistance" (Stein 2018, 139).

What does all this mean, what does this push for inclusion inextricably commingled with the pull of exclusion represent as we think and feel through the *how* of politics? In reading the university, I am brought back again and again to Saidiya Hartman's words in her text *Scenes of Subjection: Terror, Slavery, and Self-making in Nineteenth-Century America* where she writes about what she calls "the precariousness of empathy and the uncertain line between witness and spectator" (1997, 4). Hartman painstakingly recalls the obscenity of displayed, tortured Black bodies during the brutality of slavery and asks, "how does one give expression to these outrages without exacerbating the *indifference* of suffering that is the consequence of the benumbing spectacle or contend with the narcissistic identification that obliterates the other or the prurience that too often is the response to such displays" [emphasis mine] (1997, 4)? In reading Hartman's work, it is the word indifference that sticks out to me as I visualize the ways in which white men might have laughed and joked as they whipped and beat and forced enslaved people to work, dance and sing. According to the Oxford dictionary (2004) definition, indifference is to show lack of interest, concern, or sympathy. Merriam-Webster similarly defines it as an absence of compulsion to or toward one thing or another. Hartman describes in vivid detail the lack of affective compulsion toward the enslaved during enslavement, as the whip and other means were used to beat them into submission. Vicious force compelled the enslaved to stay in line. Although we have historical records of the enslaved naming their pain, it is still impossible for us to imagine everything that they must have felt. In the words of Walter D. Mignolo, "one does not see and feel capitalism in the same way across time and space and thus across different colonial settings. Instead, what you see and feel from different and differential colonial places is the colonial matrix of power ... In the colonial matrix, the legitimizing discourse encompasses authority, gender and sexuality, knowledge and subjectivity, authority and economic organization" (2015, 115). What we have learned from history is that technologies of discipline are not static – they change over time. Whips are no longer used as a form of discipline, but they evolve into other disciplinary methods of "the coerced enactment of indifference and the orchestration of diversions" (Hartman 1997, 38).

I include in these technologies of discipline "evidentiary technologies ... The same evidence can be stunningly successful at one moment in time, and fall on deaf ears at another. This accounts at least in part for the mercurial career of anthropometry in the twentieth century: the measurements may not have changed, but the audience willing to believe that skull size has a proportional relationship to intelligence certainly has" (Chinn 2000, 22). Old and new iterations of technologies of discipline and oppression serve the same function, they invoke the bodies of the predesignated non-human, non-person as evidence against themselves as a means of maintaining status quo economic dominance. We know that "pedagogies of nationhood, race, gender, sexuality, class, and culture within the imperial nation are fundamentally intertwined with the interests of neoliberal capital and the possibilities of economic dominance" (Chatterjee and Maira 2014, 7). This key set of concepts points to how the academic community appears to be transformed with changes in societal values. Thus today, we exist within a time where it is frowned upon to be explicitly bigoted and vocally racist, to engage in racist scientific experimentation – where science is supposed to be unbiased and neutral – and a number of systems have been put in place to make sure of this, for example, internal review boards. On the face of it, universities are now interested in types of measurements that appear to be progressive compared with those of the past, particularly in how they incorporate the study of race and gender – Black studies and women's studies are now part of the academy. And yet today, just as biological deterministic theories and eugenics discourse and practices were crucial to the development and sustaining of the enlightenment project and its definition of rational man and concepts of humanness (Ansfield 2015, 131), so too are the contemporary notions of inclusion, such that "the racial underpinnings of scientific knowledge and the application of this knowledge of black bodies ... always already subjugate and/exclude marginalized communities, thus bifurcating our *analytical approaches* to race, science, knowledge, and collaboration" (McKittrick 2015, 149). But I digress. Today, we now find that institutional politics, particularly around inclusion, remain tethered to the academy's racist sexist traditions of knowledge building even if this is not made obvious. The contemporary inclusionary politics of the academy is based on what Guillory calls "*indifferent* or redundant scholarship" [emphasis mine] (2012, 7) which have the result of "validating some forms of knowledge and disallowing others. ... these ostensibly neutral criteria not only regulate *what* gets said but, in making the University an inhospitable place ... also determines *who* can say it" (Hong 2008, 104).

Returning to Hartman and the concept of indifference, it seems to me that indifference becomes a critical concept which must be addressed when engaging with *how* we think about academic inclusion today. Hartman writes that we need to think about the indifference which produced a dualism. As we regard what it meant for the enslaved to be made to sing and dance and engage in "a range of everyday acts, seemingly self-directed but actually induced by the owner, [these behaviors] are the privileged expression of this consenting agency" (1997, 54). The consenting legacy is the institution itself. We need to think about inclusion as not only a matter of those being included but also about those extending the inclusion. The institutions and affiliated persons who engage in the act of inclusion will use the concept of inclusion as a disciplinary technology, making the case (like the slave master) that it is in our very nature to behave in ways which our bodies are forced to submit to, no longer by the whip, but through the often tacit "dances" of the inclusion imperative. Perhaps not by force, but by other modes of compulsion racialized and gendered bodies are made spectacles held up as contented, when they are in fact only contending. It was never about how good the enslaved were at dancing or singing, just as today successful inclusion is not so much about being a good student or otherwise displaying merit, rather it is about "political conformity," per Edward Said, who asserts: "the general atmosphere … [is] both conspiratorial and… repressive … [and] the University has come to represent not freedom but accommodation, not brilliance and daring but caution and fear, not the advancement of knowledge but self-preservation" (1991, 7). This self-preservation in the time of movements for social justice, requires that the university invoke language of inclusion as a symbol of organic cooperation and reciprocity. Similar to the slave master's whip, inclusion promotes the appearance of the happy, smiling students who are eager to "step it up lively."

Put bluntly, we cannot consider the university without also considering the plantation. McKittrick writes that the plantation "propell[ed] an economic structure that would underpin town and industry development in the Americas" (2013, 8). The whip and the academy were/are just two technologies of discipline in this development. When we consider Davies' (quoting Wynter) proposition that "education… was the 'chief agent of indoctrination by which the colonized black internalizes the standards of the colonizer other'" (2015, 218) my point becomes clear. Indoctrination works both to reinforce white, colonist hegemony while also providing a pathway to citizenship. According to Cedric Robinson, "[s]omewhat paradoxically, the more that Africans and their descendants assimilate cultural materials from colonial society, the less human they became in the minds of the colonists" (2000, 119). To McKittrick's point, the plantation is an institution that society today, including the academy, would rather keep forgotten as backward and indefensible. The academy would like to move on from its apparent involvement in propagating and propping up a colonial system, and focus on what it is doing now to "right" its hand in the wrongdoings, specifically through diversity and inclusion performances. But, as McKittrick writes, "the plantation moves through time, a cloaked anachronism, that calls forth prison, the city, and so forth. ... a persistent but ugly blueprint of our present spatial organization" (2013, 9). I will continue to demonstrate how the spirit of the plantation lives on in the university.

As the plantation was, the university is still tethered to the marketplace (Hall 2017, 270). "Colonialism's multiple projects ... cannot be seen simply in terms of having been past .... Not only is there no fixed past, but various technologies of timekeeping and various narrations of time can exist within the same temporality" (Alexander 2005, 191). In other words, according to Spillers, "[e]ven though the captive flesh/body has been 'liberated,' ... the ruling episteme that releases the dynamics of naming and valuation remains grounded in the originating metaphors of captivity and mutilation, so that it is as if neither time nor history, nor historiography and its topics, shows movement, as the human subject is "murdered" over and over again by the passions of a bloodless and anonymous archaism, showing itself in endless disguise" (1987, 68). The passing of time has not altered the materiality and character of the colonial system within which the university is implicated, neither has time changed the underlying indifference through which the oppressive and apathetic methodology of science uses the body against itself. Time has only recontextualized the delivery and dissemination of oppressive power, such that the covert recapitulation and disavowal of past racist and sexist technologies of oppression have led to the development of different types of technologies and ways in which bodies are assessed, accorded value, and managed. Today, the management of the non-person, non-human body within the university, occurs principally through the implementation of inclusion initiatives. In "And Not a Shot was Fired," Stuart Hall makes it clear that at the very time of inclusion and diversity, "when less well-prepared students from educationally-disadvantaged groups are coming into higher education for the first time ... the new echelon of managers ... restore... and restructure the institutions along market lines. They operate at the coal face of the system, and are responsible for the day-to-day implantation of the new habits, routines, disciplines, the language of calculation, the fine tactical decisions, which gradually *institutionalise* the new 'regime'. ... they speak that metallic discourse of 'managerialism'" (2017, 271).

The point I have been making is, the university has not changed its values, rather only superficially altered its original statements of mission in a way that doesn't truly transform power relations. To manage the "crisis" provoked by the student activism in the late 1900s, the university incorporated a diversity and inclusion paradigm, not derived from any form of deep, thoughtful, ethical revision of its racists and sexist underpinnings but according to Kehinde Andrews, "opened up the sector to women, the working-classes and ethnic minorities ... in a manner that entrenched rather than challenge them" (2018, 276). Although today some may think of the university as mostly a neutral, benevolent space,

> in the U.S. context its ethical-political possibilities are consistently (re) shaped by its structural entanglements with state power, the imperatives of capital accumulation, and their ordering logics of racialized de/valuation. [It] has adjusted over time to remain aligned with, and to help manage, shifting priorities of the state-capital articulation—as indeed it has done with the most recent shifts.... current crises of the U.S. university are not entirely the result of novel transformations or the betrayal of its underlying values, but rather their fulfillment. In other words ... the University is not broken—it was built this way (Stein 2018, 131).

The "crisis" mentioned by Stein is therefore, a conflict between professed ethical change and the reality of its lack. It relates to the politics of indifference and leads me back to Hartman again.

The indifference Hartman writes about remains even more imbedded today. It is most noticeable now as the calls for justice from those who have been historically denied humanity and personhood have become louder and louder. Those yearning to be let into the academy are still demanding a shift in the power structure to include justice and not merely admission. However, framing the question as one of "being let in" suggests that greater attention be placed to understanding the space and place one is being "let into" and its spatiotemporal significance as part and parcel of a new iteration of plantation logics. To be given access to a particular institution requires that we think about why one is let in, on what terms, and what is in it for the letters-in. These questions are important when we think of the pushback by Nance, the parent mentioned in the Introduction, around who is meant by "us," and when we juxtapose Nance's complaint with the 2019 university admissions scandal which exposed that rich parents were buying their children's way into top tier universities in the United States (Taylor, 2022). To be let in for affluent, white students harkens back to the Lockean philosophy of gentry education, i.e., these spaces are made for them, their admission is "supposed to be" a given. And yet for those historically without money, personhood or humanity, to be let in requires something much more: there are, as Ahmed has pointed out, conditions to this hospitality.

The conditions of being let in are manifold, borne by those desirous of inclusion, and by communities who are (un)intended collateral damage. Let me address the communities first. Over the past few decades universities have eaten up vast amounts of land around them, driving up prices. The excuse is that they do this to make the university more accessible and a place where the community can feel welcomed. But who is "the community" and what does "accessibility" even mean when, the university's neighbors are priced out and dispossessed – meanwhile the university rehearses empty apologetics after all is said and done? Sharon Stein makes an explicit call-out:

> Recent university apologies for participation in Black enslavement (such as those offered by Harvard, Brown, the University of Virginia, and Georgetown) and, more rarely, Indigenous genocide (at Northwestern and the University of Denver), signal a tentative openness to accountability. Yet, the almost total absence of subsequent actions taken by these institutions to enact redress by returning lands, resources, or other institutional wealth that was generated through this violence, or by addressing *ongoing* material and epistemological violence against communities of color and Indigenous communities on and around campus, signals a firm limitation to universities' conceptualization of accountability and to the possibility of transforming the institutional conditions and logics that produce(d) that violence in the first place (2018, 136).

Regarding those previously excluded and now welcomed, the conditions under which they are let in must be read through a lens which examines the violent nature of inclusion policies upon those in the category of the previously excluded. I want to explore here the ways in which the politics of indifference is propelled by university cooptation of the desire of the pervious excluded for real inclusion, what I call the promise and politics of yearning.

The politics of indifference and the promise and politics of yearning

As Chinn states, it is easy for most of us "to fall into the pattern of looking at bodies and believing we know something – all – about the people who live their lives within them ... to distance ourselves from the embodiedness that dehumanizes" (2000, 169). In this particular moment when the fight for the body is a fight for power, it is important that we understand how this fight is being fought and won at the level of its construction as evidence of its essence – legible, visible proof of its truth – and how the body as proof, as demonstrated in the previous chapter, is built by and into scientific evidence from the late nineteenth through to the late twentieth century. Meanwhile, one cannot miss how this fight is also happening in the realm of the affective. As mentioned above, the persistent and prolonged call for justice has presented a situation which prior to the mid to late twentieth century was perceived by those with power as unimportant, or as creating a problem which needed to be quickly identified and quelled. According to Chinn, "[o]nce the problem [is] "identified" ... a discourse of measurement grows up around it, defines it, and constructs a body of knowledge around it, which is then wielded by experts to contain what was previously amorphous and uncategorizable. Simultaneously, these discourses of measurement displace previous ways of knowing about the "problem" before it was defined as a problem: when it was simply an arrangement of events, feelings, ideas, or reactions" (2000, 16). History has demonstrated that the ability to measure and categorize in order to manage and control is crucial to power's stability and resiliency.

Our current mode of registering the body as an evidentiary instrument in order to shore up the equilibrium of power in the academy is diversity and inclusion. Many of us who desire justice and equity, and who do the intellectual and activist work to pursue these goals, have been incorporated into the university as bodies counted and measured as "diversity." To this end, we are important to the institution only insofar as our raced and gendered bodies provide the numbers required to maintain an unchanged balance of power by a performance: showing themselves to be responsive and good at working toward change. Meanwhile, our presence also provides the university with the *bodies of knowledge* that continue to interpolate our bodies as evidence and to provide the university with "knowability" (Chinn 2000, 10). The stickiness of it all is that the university uses gestures of inclusion as a signal to all about its benevolence, presenting itself as a welcoming, caring place, invested in the success of all, the futurity of everything that is good about society, and "as a sign of the University's 'excellence'" (Hong 2008, 106) such that, who would not *want* to go there? The point is that the university, this bastion of rationality and analytics, needs/wants/desires the bodies of the marginalized and minoritized, as they/we also yearn for a chance to be a part of the university. This yearning, as it is caught up in the everyday spatio-temporal organization of our society, is therefore political, as it promises to both parties something that they each really want. There is no doubt that touting diversity goals and statistics makes the university look good. The more Black, Brown, female, queer, working-class, international, disabled bodies the university recruits to fill its quota, the more it can hold on to its claim of excellence. And yet, riffing off of Hartman again, being enumerated in this way means these bodies are yoked to "will[s], whims, and exploits ... and by the constancy of ... unmet yearnings" (1997, 49) which in the case of the enslaved ranged broadly from food to freedom, and today to the possibilities one imagines from gaining an education. Hartman's point here is that our most profound desires, are "ensnared in a web of domination, accumulation, abjection resignation, and possibility. [This web of domination is] nothing if not cunning, mercurial, treacherous, and *indifferently* complicit with quite divergent desires and aspirations, ranging from the instrumental aims of slave-owner designs for mastery to the promise and possibility of releasing or redressing the pained constraints of the captive body. It is the ambivalence of pleasure and its complicity with dominative strategies of subjection" [emphasis mine] (1997, 49–50). Applied to our present context of educational attainment, Hartman's analysis demonstrates how our yearning becomes political, as the promise of this desired education is wrapped up with the competing and conflicting ones of the university such that to achieve these desires we are forced to be complicit in our own oppression.

As it did in the nineteenth and twentieth centuries, the body still shows up today as evidence – only now, an evidence of diversity – used for its evidentiary value to shore up institutional power. What we learn from Hartman is that we need to complicate and thus reveal the problematic of how technologies of power work, that is, what does singing and dancing mean, both/and as a desire for relief, and as by force of the lash? What does it mean to yearn for a place in the university, both/and to be desired by the university for a very specific purpose and habitation? What types of relief (on all sides) come from this habitation, what types of disappointments, what types of trepidation? Like the enslaver, the university requires the body to do work to produce its desired capitalist outcomes, but also by constructing that body in very precise ideological ways, it expects said bodies to perform in prescribed ways for the university's sole benefit, and in addition said constructions are to be used as evidence against (as well as of) these bodies to tell us something about them: their deficiencies, whether inherent or derived from their cultures and/or communities, in order to justify not only why they need the university (which is doing a benevolent thing by letting them in), but also to demonstrate why in turn the university needs to use their bodies as evidence to prove the impact of its benevolence. These terrible entangled dualities and covert exchanges exist – the quantifying of bodies which are more "needed" than perhaps "wanted," and the cruelty of this bargain is doubled in that it is covert. Only the promise is offered, but the demands are left a secret to be discovered but never admitted to. To get to this place today, where the body is displayed as evidence justifying the "legitimate" perpetuation of power requires, in my opinion, inordinate amounts of indifference. An indifference not only to the impacts and outcomes of measurements on physical and affective bodies, but also indifference in how "the desire to find in the body proof of 'difference'" (Chinn 2000, 6) needed to implement university diversity and inclusion policies, in fact uses the body to generate this evidence of difference it seeks to find.

From this perspective then, desire/yearning becomes crucial to how the university, despite its wanting us to believe otherwise, is an exploitative affective space! The university uses its and our yearning as a political shaping tool for the purpose of propping up power through the use of scientific measurement as justification. As we yearn to exist in the very place where the body has been picked apart literally and figuratively in the name of scientific investment for the good of humanity (read, white and male), understanding how the politics of indifference operates in the academy becomes important to how we think about the ways in which bodies are used – measured – as evidence of how hard the university is working to repudiate its colonialist, racist and sexist roots through a process of diversity and inclusion, welcoming bodies – of people and knowledge – that embody, by virtue of their/our existence in this space and time, visible, legible, empirical evidence of the university's truth-claim that it is working toward justice.

The university's deployment of a politics of indifference is not confined to this contemporary period alone. It has historically engaged and depended on this politics to mobilize the racist science it has used in order to produce the measurements of the nineteenth and twentieth centuries described in Chapter 1. Consider how indifferent one had to be about using Black people as data by disturbing the sanctity of the resting places of Black people, digging up their bodies and dissecting them as Diana Ramey Berry reminds us that the academy did. Both Berry and Wilder write of how in the nineteenth century, "[t]he faculty and students harvested colored corpses from the African cemetery for years, dragging cadavers across Broadway [in New York City] to the dissecting table" (Wilder 2013, 198). Or consider the indifference in the "case of Henrietta Lacks, whose cancerous cells were taken without her consent, confirm[ing] … that [Black] bodies and body parts" (Berry 2017, 193) were indispensable to the university even as they were excluded from intellectual life. According to Berry, "[t]he merits of a college could be reasonably measured by its collection of human remains, a good catalog of skulls, skeletons, and skins being a considerable advantage in a competitive academic market" (2017, 193). The academy's reliance on indifference, and lacking the will to truly understand persons/bodies outside of the evidentiary value and utility that the dehumanized and depersonalized body provides has meant "[t]he language of blood, skin, and bodies have proved to be amazingly mobile and adaptable to any number of agendas" (Chinn 2000, 22). The above examples of university indifference are just predecessors of the current iteration of university applied indifference pervasive in our society, specifically as demonstrated through the actions of university administration and academics within this contemporary moment of measurement for inclusion, diversity and the appearance of equity. "Modern/imperial epistemology," even in its diversity, is still always imperial (Mignolo 2015, 116).

M. Jacqui Alexander posits "a certain politic of disingenuity … disingenuous bureaucracy" (2005, 128) which pervades the university such that we must be wary of its calculated dishonesty. I argue that while this is so, what the academy does today goes beyond disingenuous bureaucracy. Like Denise Ferreira da Silva, who writes about "the ethical indifference with which racial violence is met," it is the "determinacy, along with separability and sequentiality [that] constitutes the triad sustaining modern thought, [which] operates in the ethical syntax in which this *indifference* makes sense as a (common and public) moral stance" [emphasis mine] (2017, 4) and which distinguishes it from mere disingenuousness. It is in this way that I have come to think of indifference as a common and public stance as political; a politics of indifference, in that it organizes our behavior. As I mentioned briefly in my Introduction, the practices of the politics of indifference in our contemporary society work to displace and use bodies for its own purposes by deploying a so-called objective science/academic investigation/reason, developed for the good of society. Hence, I define the politics of indifference as the seemingly detached, apathetic, reasoned and systematic policy approaches used by institutions in society surrounding considerations of gender and race inclusion and exclusion, and how these policies lead to sanctioned practices which have traumatic consequences for women and Black and Brown people. The politics of indifference is premised on a colonial/patriarchal (mis)reading which informs our societies at various points in history, including our contemporary societies, and is concerned with the normalizing of bodies through an attention to difference. This attention to difference is never in the interest of those considered different – in fact there exists a deep disregard, the violence with which is concealed by a façade of objective scientific/rational/academic investigation which institutions claim to be in the best interest of society. It cannot be forgotten that this is a capitalist society where certain bodies have long been commoditized and assigned value as objects of exchange and laboring property. The legacy of this system of commoditization and valuation applies to how some bodies are included in the university today: laboring to be included, even as their bodies are assigned based on how much value they provide for the university, which remains indifferent to the negative impacts of this type of inclusion.

I want to very briefly make a distinction between the politics of indifference and the racism and sexism which lead to the performance of racist and sexist science. I am in no way suggesting we replace our understanding of racism and sexism with the politics of indifference. Racist and sexist science and the harmful measurements that scientists have performed at academic institutions do stem from colonial racist and sexist ideology. However, the reason why it is important to make a distinction is to demonstrate how the politics of indifference is crucial to the ongoing mechanisms of racism and sexism. It is the politics behind the ways in which white institutions organized, operationalized, and justified their racist and sexist scientific practices. To operationalize its racist work, racism depends upon, among other things, the cooperation of those people and institutions (explicitly racist and not) who, and which, willfully and intentionally perform tasks knowing that the outcomes will have impacts that are harmful and violent to the marginalized Others, Black and Brown people and women, but are indifferent to what that outcome is. By indifferent I mean, they do not/or choose not to care/feel for the persons upon whose bodies the outcome of their actions will impact. The politics of indifference is calculated: it callously operates to be socially and economically beneficial to people and institutions who wield it, as they perform the work of capitalism. The calculations are never primarily about the people who are the subject of measurement. It is important to tease out the politics of indifference as a component of the operationalization of racist measurements because this is how we may come to understand and *read* the ways in which affect belongs to a system that is normally read as rational.

The power of the politics of indifference is evident in the ways in which members of the academic community (including those who truly believe in doing no harm, even doing good through their own work) themselves go along with, or are forced to defend, uphold and protect the institution by their silence, using the veneer of academic freedom and rigor. In 1917, Randolph S. Bourne wrote in response to the termination of two Columbia University faculty members, James McKeen Cattell and Henry Wadsworth Longfellow Dana for their public opposition to World War I (which also led to the resignation of renowned historian Charles Beard in protest): "The University produces learning instead of steel or rubber. ... As directors in this corporation of learning, trustees seem to regard themselves primarily as guardians of invested capital. They manage as a sacred trust the various bequests, gifts, endowments which have been made to the University by men and women of the same orthodoxies as themselves. Their obligation is to see that the quality of the commodity which the University produces is such as to seem reputable to the class which they represent. ... the reputation of a university is comparable to the standing of a corporation's securities on the street" (1999, 152–153). So-called radically political academics, members of the knowledge production class removed in many ways from the implications and impacts of their production, have traditionally been protected by the university.[1] Whether their politics places them at odds with the institution or squarely in its center, academics benefit by invoking notions of respect and freedom of thought, and can rely either on the valorization of their actions by their peers on the one hand, or their silence on the other. Wanting to have it both ways – the freedom to remain objective and yet politically active and engaged – academics act as trustees of the institution as they either argue for the freedoms as guaranteed therefrom or sit in silence. They are complicit in the university's indifference.

According to Sheila Slaughter and Gary Rhoades, "[p]rofessors and higher education interests certainly served the industrial economy and the state, but in doing so they arguably gained some power by claiming a social contract with society in return for disinterested, nonpartisan research ... In the information society, knowledge is raw material to be converted to products, processes, or service. Because universities are seen as a major source of ... knowledge, they are in the process of establishing new relations with the global economy. ... The new economy treats advanced knowledge as raw material that can be claimed through legal services, owned, and marketed as products or services" (2004, 15). Producing this raw material requires data and measurements, and invariably bodies themselves as means of producing data and measurement, such that for their own profitability and notoriety, universities are indifferent to *how* their apathetic use of power and indifference has real, affective and material consequences for the already minoritized and marginalized. Quoting Slaughter and Rhoades again, the university incorporates, and quantifies and abstracts students as "'inputs" and 'outputs' in order to "strengthen the market positions of the colleges or universities that enroll them ... and signals the success of the school [thus] ... fashion[ing] a virtual cycle of competition in which students and institutions in the same (elite) market segments compete ever more vigorously with and for each other, contributing to the instantiation of an academic capitalism knowledge/learning regime" (2004, 43–44). In this academic market place bodies are incorporated for their use value.

I also want to make a distinction between the politics of indifference used to frame this project and Iris Marion Young's politics of (positional) difference (2005, 6). Young argues that "public and private institutional policies and practices that interpret equality as requiring being blind to group differences are not likely to undermine persistent structural group differences and often reinforce them," and instead argues that we must "explicitly recognize group difference and either compensate for disadvantage, revalue some attributes, positions or actions, or take special steps to meet needs of and empower members of disadvantaged groups" (2005, 6). Young makes the point that institutions must face their lack of understanding and fears about group difference, and as oppressed groups take hold of the reigns of naming, defining and empowering themselves, their difference should not be viewed as otherness but rather as relational similarities and dissimilarities. Young argues, as I have throughout this chapter, that there is no such thing as impartial, unbiased, value-neutral science (1990, 193). Similarly, she points out it is imperative that institutions examine their policies, and practices and voluntarily address the ways in which they perpetuate inequities in their structures, so as to further social change (1990, 39). I argue while well-intentioned, Young's politics of difference is insufficient to addressing the problem of diversity and inclusion as it is implemented through the politics of indifference. We need to look deeper.

The distinction between the politics of indifference and the politics of positional difference is that while the politics of difference "worry about the domination some groups are able to exercise over public meaning in ways that limit the freedom or curtail opportunity... [and] challenge difference-blind public principle" (Young 2005, 28) or the ways in which domination manifests, what I am calling the *what* of domination, the politics of indifference goes deeper to examine the *how* of domination, that is, how does this politics come to be, what is behind its operationalization, the methodology behind it? In thinking through the how of politics, I think about the apathy inherent in indifference such that it underscores its methodical application. This is what I call the apathetic methodology of power, used when enacting policy decisions about the included. In her critique of the politics of difference, Ien Ang states, "idealized unity is a central motif behind a politics of difference which confines itself to repairing ... friction .... The trouble is that such reparation strategies often end up appropriating the other rather than fully confronting the incommensurability of the difference involved. This ... only reinforces the security of the white point of view as the point of reference from which the other is made same, as symbolic annihilation of otherness which is all the more pernicious precisely because it occurs in the context of a claimed solidarity with the other" (2001, 397). Ang's quote sums up why simply looking at addressing the ways in which difference is manifested in terms of how it shows up on our campuses is insufficient and reinscribes otherness. As such, we need to address the means by and through which these unequal conditions come to exist and address those if there is to be any semblance of change.

When seen through the lens of the politics of indifference, it is clear that even when, as Young suggests, group difference is recognized and revalued, the apathetic methodology of power operates to further disenfranchise and marginalize groups considered different *while appearing not to*. The reason why this is an important distinction to make, especially in the academy, is because as *difference* becomes accommodated through diversity and inclusion programs, students and faculty exist in this accommodation *in and by virtue of* difference. In other words, difference is given as the reason the marginalized are accommodated. Young writes, "the goal of achieving greater justice legitimates preferential treatment" (1990, 199). However, the language of preferential treatment can act to further marginalize in two related ways. One, any accommodations or re-evaluation extended through the extension of preferential treatment, will work to shore up power's claim to benevolence even as it continues to use those bodies for its own gains, which the university can further capitalize on in the form of pushback when, two, this perceived preferential treatment results in expressed resentment by the historically assumed "us," like Nance mentioned in the introduction. Both of these scenarios work to place the "not us" in danger.

This double bind is evident as the institutions make accommodations for Black, Brown and women students and hire more Black and Brown faculty and staff. The administrative assumption is that they have done their part for the good of diversity through the implementation of their policies, and nothing more could reasonably be demanded. These institutions often display an indifference, if not outright irritation, to requests from faculty, staff and marginalized and minoritized students who advocate for additional change and resources, and the politics of indifference is further perpetuated. Young also argues in her work that those who are othered would rather "prefer a stance of respectful distance in which whites acknowledge that they cannot reverse perspectives ... and thus must listen carefully across the distance" (1997, 345). I, however agree with Ahmed that this approach is shortsighted, and that we must go beyond Young's work of listening, to pursue *reading* – closely. There is, according to Ahmed, "a danger in assuming proximity *or* distance ... as ...[a] point of entry, [as it] fails to recognise the implication of the self in the encounter, and the responsibility the self has for the other to whom one is listening" (2000, 157). One needs to be able to closely read the encounter and the context of the encounter as a form of practice. Developing a reading practice allows us to see the signs of oppression in the way we show up as the university's evidence of diversity, using our "difference" as part of "a system of valuation in which the dominant is rendered invisible and the subordinate hypervisible for the purposes of control, and the reverse for the purpose of normalization" (Chinn 2000, 8).

Reading helps us to consider what Ahmed quoting Robert Stam refers to as "the relationship between mouths and ears in the communication of injustice ... [as] 'the circuit for mouth to ear is also an *a priori* open or public thoroughfare, the message sent along it take the form not so much of a sealed and esoteric letter but as a postcard for all to *read*' (1995: 78)" [emphasis mine] (2000, 157–158). Because we have been incorporated into the university, in which encounters are premised on historical pronouncements, listening is not a neutral act – particularly with regard to listening for evidence of truth. According to Chinn quoting Bakhtin, "Once a phenomenon has been shaped through expectations that it represents some kind of evidence, it 'cannot fail to brush up against thousands of living dialogic threads, woven by socio-ideological consciousness ...; it cannot fail to become an active participant in social dialogue' (Bakhtin, 1981: 276)" (2000, 10). So that the question which emerges is, what power dynamics take part in this dialogue to predict who listens, how, and to whom?

Therefore listening and speaking are social encounters that are written upon a history of injustice about who knows and can tell the truth (Ahmed 2000, 158). As bodies previously excluded from "truth-telling" are now entering into the place of "truth," when these bodies communicate injustice how then do those with power listen, hear and respond, particularly from a distance? How does power respond from a distance except through a rhetoric of evidence which "is mute unless it speaks the language of rules," that is, "in the language of normalization through either a discourse of common sense or the imprimatur of professions and experts" (Chinn 2000, 19, 21)? According to Hall, cited in Grossberg, "[i]t is not the individual elements of a discourse that have political or ideological connotations, it is the ways those elements are organized together in a new discursive formation" (1986, 55).

Davies sums up this observation: "Speech, then, is as much an issue of audience receptivity, the fundamentals of listening, as it is of articulation" (1994, 21). Because the university listens and communicates from a distance, they continue to engage in ways that are harmful. They do not (or pretend not to) hear the rise and fall in the timbre of our voices, the quiver, the fear, the sadness, the joy and celebration. We speak across the distance they create, and they respond in the language of rules, policies and procedures, committees, task forces, tribunals. They often hear and address the what (the thing) of the injustice without looking behind/beyond the injustice to read – "the arrangement of structures that fix us in relation to each other, and is represented through narratives that allow us to understand those relations discursively" (Chinn 2000, 8). They do not get the importance of affective, close and passionate communication which considers the positionality of the speaker.

According to Puwar, people inhabit "a very particular speaking position; the utterances of these people are linked to their bodily existence. Their voices are anchored to what they are seen to embody. This is a burden and a connection that is not the first consideration that comes to mind when a white male body speaks, writes or creates. He just speaks as a human, because race and gender are ex-nominated from his bodily representation. While we can no doubt show how this universal figure of a human, who is commonly assumed to be speaking from nowhere is speaking from somewhere, as an embodied being (in terms of nationality, gender and class, for instance), he nevertheless occupies a position of privilege of invisibility" (2004, 72–73). Puwar's point here definitely gets to the point of understanding embodiment as being part and parcel of the utterances of not only the mind but also the soul. As the white, male body is not overwritten with the same language of the racial and sexual analytics as Black, Brown and women's bodies. Imani Perry's writing about how passionate utterances, as opposed to listening from a distance, are intimate gestures toward being together as humans, can help us to think about the ways in which nonauthoritative and more open communication can act as avenues to moving pass and abandoning so-called truth claims about others. Perry writes, "[b]eing open to passionate utterances is about the possibility of being moved... To heed the passionate utterance is to witness in ways that are far more fundamental than agreeing to believe in a set of doctrines ... It is to allow for a resonance beyond the familiar that will shape our ways of being and seeing. This is requisite for the liberatory political imagination to blossom, for us to begin our revolutions" (2018, 224–225).

This is how the politics of indifference works: through listening at a distance. Listening at a distance is how what I refer to as the apathetic methodology of power works at institutions. Listening from a distance as deployed by the apathetic methodology of power can result in words falling on deaf ears, so to speak: this dynamic looks like requests for the minoritized and marginalized to speak their truth and bare their souls about, for example, how the COVID-19 pandemic is uniquely, negatively impacting them as a group, and then being told by the university simply that there are no resources to help but that they are in the midst of studying how the virus is impacting the marginalized. The university offers these words while at the same time it holds wages steady, does not provide benefits for contingent faculty, and either does not provide or provides subpar insurance for graduate student workers during a pandemic. All the while, these marginalized people are depressed, homeschooling children, financially helping families who have lost jobs and much more. Intentionally blind and deaf to the additional stresses which come from the isolation and stress of being marginalized, maybe the administrative assumption is "things are tough all over, why do 'these people' need 'extras'? Meanwhile these bodies offer the university coveted "diversity points" which makes them "useful," and additionally their admission to the academy is taken to be an "extra" privilege in itself, and so they should be quietly grateful. These bodies serve their purpose to the university, and the "welcome to our community" ends there. Difference is merely a symbol or a game token, and the meaning of difference (which is that there is a material and affective difference of circumstance) is ignored. The indifference to a recruited population's wellbeing, mental health, and general ability to live and pay bills is appalling. It reminds me of the tragic case of Thea Hunter, whose death was mentioned in the Introduction to this book (Harris 2019). In her death, I see clearly how the indifference surrounding "how the epistemologies organizing the University might manage racialized and gendered bodies to the point of exhaustion, breakdown, and death" (Hong 2008, 105).

The apathetic methodology of power is a commitment by those in decision-making positions to hold up science, metrics, and academic rigor while dispassionately implementing policies and practices, unconcerned with how said policies affect the real-life experiences of the marginalized and minoritized. It is part and parcel of the legacy of what Denise Ferreira da Silva calls "the *analytics of raciality,* as a productive symbolic regimen that institutes human difference as an effect of the play of universal reason" (2007, 3). Just as "the writing of modern subjects in the post-Enlightenment period would also require the deployment of scientific tools, strategies of symbolic engulfment that transform bodily and social configurations into expressions of how universal reason produces human difference " (da Silva 2007, 3–4) so too our current diversity and inclusion policies also require it.

In the following chapter, I will illustrate how indifference works as a political mechanism on many levels. First, I will demonstrate how institutions focused on upholding the perceived societal norms and agreed-upon institutional projects at this contemporary moment still value modes of scientific/academic expertise that rationalize their support for the adoption and implementation of norms that allow them to be indifferent to the outcomes of their actions, and which thus absolve them from further action – whether that be to repudiate action(s) or withdraw, or provide further or other types of support beyond the letter of policy and procedure, "best practices," and in so doing, absolve the institution from any and all blame for the damage done. Second, as mentioned in the Introduction, because I am interested in how this politics of indifference is focused on those bodies considered to be the embodiment of difference specifically within academic spaces, I examine how these bodies exist *as* and *in* difference, that is, marked as different while navigating the space of difference (read, as a space that is good or even exceptional at diversity) as the body that marks the space as such. In essence, I am asking how does difference exist *as* difference *in* difference (read diversity)? How do these differenced-bodies navigate difference as relegated mostly or solely as that difference, confined to participate in the ways that highlight this difference? Finally, I demonstrate what my research has shown: that those marked by, in/difference – the politics and the perception – have been forced to navigate these spaces *in* difference by conforming to and/or pushing back against said indifference. I am therefore interested in the negotiation that occurs between universities as they engage with the politics of indifference, and with differenced-bodies in difference, and the consequences of this negotiation, including: how this experience is felt, and how do those that provide this difference feel the effects of the ways in which they are incorporated – that is, how they feel the university or how the university is felt by them. If, as Ahmed argues, diversity can function as a containment strategy (2012, 53), how does it feel to be contained?

The significance of affectivity, understanding how the university is felt should not be underestimated. As the university is a place held up as the bastion of reason and rationality, there is an obvious "irrationality" that comes to the fore for me as I think about the logistics of the university and the ways in which, even during a pandemic and the exponential increase in tuition, people keep flocking to its gates. While there are those willing to put up with the institution's indifference to paying a living wage, the impacts on the neighborhoods, and failure to provide proper healthcare, those variables are part of the calculation made to maintain the politics of indifference; as the university can be certain that within the society in which we live, there exists another component which is important to both the institution and those whom it needs to attract/include in order to continue being successful, and that leads us back to the idea of yearning. And an exploitative economics of yearning! That is to say, "[y]oung people are driven to college by a desire to learn ... This flow is unchecked, and it is what makes the higher education market inelastic. Prices skyrocket, and the customers continue to throng at the door. There is little choice when the job offerings are fewer. As jobless growth overcomes the economy, and as nontradable services are the only boom sector in the U.S. job market" (Prashad 2014, 332). The university provides folks with what Berlant calls "fantasies of the good life" (2011) based on a "romantic story of education ... portray[ing] the student ... as engaging in a quest, climbing up education's levels, overcoming obstacles ... at each step, and rising toward an image of the good life defined by success, security, independence, maturity, and happiness" (Meyerhoff 2019, 22–23) and this desire upholds "...the University ...[as] a romantic hero" (Meyerhoff 2019, 48). But, what happens to these desires/fantasies once one gets to the university space, when this space of "life-building and expectation whose sheer volume so threatens what it has meant to 'have a life' that adjustments seem like an accomplishment" (Berlant 2011, 3)? Linking back to Hartman, how does the university as the image of freedom leave unable to discern whether we are "driven by the lash or by the inward drive," when "doubled over by the sheer weight of ... responsibilities, hopeful and obedient, work was to be its own reward, since the exertions of ... labor [are] also demonstrations of faith" (1997, 135). And so this affective relationship with the university is built on yearning and hope, for the price of obedience and faith in the institution. And yet there is a steeper price.

To say simply then that the university is a place of the mind is not to look behind the politics of indifference, to give the university a good read. The university peddles in affectivity and affective connections while also being a place where "psychic and affective negative feelings about blackness ... are implicit to a symbolic belief system of which antiblackness is constitutive ... [and] informs neurobiological and physiological drives, desires, and emotions—and negative feelings—because it underwrites a collective and normalized, racially coded, biocentric belief system wherein narratives of natural selection, and the dysselection of blackness, are cast as, and *reflexively* experienced as, commonsense" (McKittrick 2016, 83). Emotions play a critical role in the way in which the university operates. Its affective underwriting and the way in which it deploys a politics of indifference plays on the desires of the marginalized and then, according to Alexander, calls up "good intentions to account for the reproduction of inequity. ... [and] iterates its own emotional requirements: gratefulness for being rescued" (2005, 176). Alexander continues,

> [w]hen one refuses to pay the debt by seeming ungrateful, by calling for accountability, by presuming belonging, and by insisting on the right to truth and to knowledge derived from one's social historical location, the existing emotional cart is upset by the prospect of no longer being able to rely on the same terms for balance. If the cognitive mediates the relationship between knowledge and experience, the emotional is no less an important mediator. Emotional interests are simultaneously embedded within knowledge interests and in the organization and distribution of resources, both material and psychic. In part this explains why questions about the reorganization of knowledge resources are experienced as threat, loss, and displacement. But in the context of asymmetries of power, such feelings can be a simultaneous exercise on the power of privilege, particularly since power has the privilege of reasserting itself and of reconsolidating its interests, even in moments of perceived loss" (2005, 176–177).

The university knows that it can depend on students wanting a piece of the good life, especially those who come from a long line of people who have been excluded. The university politicizes this yearning with its push toward, as Paul Lauter writes, "winning the hearts and minds of young Americans to the fantasy that their interests are at one with those of [corporate] executives. Such lessons are reinforced within the multiplying classrooms devoted to promoting enterprise, marginalizing labor, submerging the realities of social-class disparity, and above all, promoting the underlying ideological tenet of free market capitalism: individualism" (2002, 200).

Despite gender and racial "mediations entrenched in capitalist inequalities in the United States, optimism involves thinking that in exchange one can achieve recognition" (Berlant 2011, 43) such that once again, yearning is used to push people toward a manufactured "belonging" to a capitalist machine. To return to the analogy of the whip vs. inward drive, what does this "belonging" imply?

A 2005 College Board study found that public and private colleges are increasingly unaffordable to U.S. students, and that for the lowest income earners, the cost of attending college, even a two-year college, can cost more than a third of their family income. "These high prices come at a time when buying power of family incomes has declined and when outright grants given to those who need it have been replaced by merit-driven (public and private) loans." Yet there seems to be no letup in the desire of young people to go to college: "in October 2005, almost 70 percent of high school graduates went to some kind of college" (Prashad 2014, 331). According to Vijay Prashad, "The cost of college tuition has risen from 23 percent of median annual incomes in 2001 to 38 percent in 2010" (2014, 331). Colleges are passing on their operational costs to students and their families, who to cover the costs of tuition and other incidentals are becoming extremely indebted.

We see how universities appeal to the affective through the manipulation of advertising and the visual language of marketing materials as they attempt "to shape the perceptions and choices of consumers in the student marketplace. Such activities do not necessarily maximize, and certainly do not prioritize, student needs" (Slaughter and Rhoades 2004, 283). They highlight their value proposition by holding out their institutions as being "more desirable educational commodities than they actually are ....mak[ing] the institution a more attractive consumption item" (Slaughter and Rhoades 2004, 284). Being accepted into these institutions as a way of achieving the good life and participating in this "economy gets confused with social belonging" – and so this manufactured fellow-feeling plays into what Berlant calls "Cruel Optimism" (2011, 186). Here I riff off of Hartman's *Scenes of Subjection,* observing how the university portrays the minoritized and marginalized as happy through various means – as smiling faces on glossy brochures, and performative celebrations such as during Black History Month, where Black folks are obliged to do more work to make the university seem exceptional through showcasing Black joy and resiliency – such that, like Hartman, I ask then what are the "dimensions of this investment in and fixation with negro enjoyment, for these encounters with the enslaved [student] grant the observer access to an illusory plentitude of fun and feeling. I contend that these scenes of enjoyment provide an opportunity for white self-reflection, or, more broadly speaking, the elasticity of blackness enables its deployment as a vehicle for exploring the human condition, although, ironically, these musings are utterly *indifferent* to the violated condition of the vessel of song" [emphasis mine] (1997, 34).

Once students accept and enroll in these institutions, the university engages in the performance, an example of which opens this text, transitioning them "from consumer to captive markets, offer[ing] them goods bearing the institutions' trademarked symbols, images, and names at university profit centers such as unions and malls" (Slaughter and Rhoades 2004, 1-2). And once these students graduate, the university continues to peddle in the affective as they appeal to the student's sense of school pride and feelings of generosity, to appeal to them in another sense, this time as alumni, to "give back."

In her 1990 text *Yearning*, bell hooks writes,

> I gathered this group of essays under the heading *Yearning* because as I look for common passions, sentiments shared by folks across race, class, gender, and sexual practice, I was struck by the depths of longing in many of us. Those without money long to find a way to get rid of the endless sense of deprivation. Those without money wonder why so much feels so meaningless and long to find the site of "meaning". ... All too often our political desire for change is seen as separate from longings and passions that consume lots of time and energy in daily life. Particularly the realm of fantasy is often seen as completely separate from politics. Yet I think of all the time black folks (especially the underclass) spend just fantasizing (12).

This passage from hooks' collection of essays is instrumental for two reasons. First, it demonstrates how as Berlant points out, "the precariat must be a fundamentally affective class, since the economic and political processes that put people there continue to structure inequalities according to locale, gender, race, histories of class and political privilege, available state resources, and skills ... in the affective imaginary of this class, adaptation to a *sense* of precarity dramatizes the situation of the present" (2011, 195). Experiences of everyday uncertainties for poor people, those who have been left out, and who sit on the margins can produce any number of conflicting emotional responses – anger, frustration, contentment, resignation, joy, hope, happiness, excitement, disappointment, etc. Those who see the university as their way out of precarious social and economic positions yearn for it, even if it means inhabiting for a time the precarious conditionality of belonging in the university. What hooks and Berlant demonstrate here, is that for the precariat there are frightening yet publicly unspeakable affective stakes to succeeding in the university.

Second, it also helps us to understand how the desires of those who have been left out as beneficiaries of the university project should be seen as inherently political. Their desires rise to the level of politics because these desires are mobilized by and against them. Being excluded from the body politic can act as a catalyst to mobilizing yearning for a better life for selves (and community). But it is precisely because of this exclusion and its potential to mobilize our yearning for a better life that we "assume our position as subjects in a world and therefore *it is in us* as a structuring condition for apprehending anything ...Our sense of reciprocity with the world as it appears, our sense of what a person should do and expect, our sense of who we are as a continuous scene of action, shape what becomes our visceral intuition about how to manage living" (Berlant 2011, 52). If our yearning is premised on our structural conditioning as subjects, then it is always already mediated externally and thereby open to cooptation and politicization, particularly when that for which we yearn has been historically denied. As Sally Davison observes,

> [i]t is not that everything is always politics. The ground keeps shifting. That is the point. The crucial issue is that any site in the social formation, in any particular moment, *can become* the condensation of political antagonisms; the site of evolving, potential political forces; and the terrain on which political allegiances are made or unmade. How this occurs, or where the terrain is to be located is a contingent matter that no formal theory of politics can stipulate or anticipate. In this sense, the place of politics is frequently *displaced*, meaning that what is significant politically may not inhabit, or only partially inhabit, the institutional arrangements of formal politics (2017, 7).

With that said, the institution which seeks to arrange its formal politics around rationality and normality actively disavows the politics of yearning because it has historically tied affectivity to the bodies of the non-human and the non-person. This is precisely why yearning in this context is political.

Quoting Berlant again, what "we see forming here [is] submission to necessity in the guise of desire; a passionate attachment to a world in which they have no controlling share; and aggression, an insistence on being proximate to the thing. If these motives stand as the promise of the scene that will provide them that holding feeling they want, the proof that it's worth investing in these forms is not too demanding" (2011, 177). To inhabit the university in this way requires a measure of disidentification, which I explain in more detail in the following chapter. In placing hooks and Berlant in conversation though, what we can tease out is the connection between cruel optimism and yearning to what looks like as a tacit agreement – maybe the right word is resignation – to embody the university's measurements of diversity as an investment in a promise of the good life.

In the words of bell hooks, we yearn to know a time after discrimination and oppression –

> about what our lives would be like if there was no racism, no white supremacy. Surely our desire for radical social change is intimately linked with the desire to experience pleasure, erotic fulfilment, and a

host of other passions. Then, on the flip side, there are many individuals with race, gender, and class privilege who are longing to see the kind of revolutionary change that will end domination and oppression even though their lives would be completely and utterly transformed. The shared space and feeling of "yearning" opens up the possibility of common ground where all these differences might meet and engage one another. It seems appropriate then to speak of this yearning (1990, 13).

hook's quote demonstrates that yearning is complex. We must understand that this yearning is *also* what often spurs people to political action, including through protest movements as a part of the quest for education and inclusion. According to Angela Davis, "The yearning for knowledge had always been there" from as early as 1789 when Prince Hall and others petitioned the state of Massachusetts for the right of Black people to attend Boston free schools, and it was this yearning which, when the petition was rejected, led him to establish his own school (1983, 101). The politics of yearning is grounded in a history of looking on, seeing, and wanting more. Black children in particular are taught that they must work hard for what they desire and that this desire can be fulfilled through schooling. The filter through which a better, more beautiful life could come was through education, and the university is held up as the ultimate path.

Receiving lessons from elders about passion and beauty and place is a story that can be told by most Black folks who were taught that they can be all that they want to be despite whatever hand life has dealt. hooks writes that she learned from her grandmother, it is *"the predicament of the heart that makes our passion real. ... She has taught me "we must learn to see"* (1990, 103) and not just listen across the distance. Like hooks, I often share my own story of how my grandmother always urged me on, and how before her death over 20 years ago, she encouraged me to reach for the stars. I still hear her voice, intimately and passionately guiding me on, and I know mine is not an isolated story. Today my mother still says to me frequently and with much love and intimacy, "G (her pet name for me), anything you do is well done," as a way of validating my feelings around whatever professional conundrum I find myself in at the time. But it is not just Black folks, as hooks writes: "Yearning is the word that best describes a common psychological state shared by many of us, cutting across boundaries of race, class, gender, and sexual practice. Specifically, in relation to the post-modernist deconstruction of 'master' narratives, the yearning that wells in the hearts and minds of those whom such narratives have silenced in the longing for critical voice" (hooks 1990, 27). This is a longing to join a desirable narrative (symbolized by the university) as one's own voice, to add the voices of those silenced ancestors and beloved family/community so we can all reach for the stars. "When we talk about an object of desire, we are really talking about a cluster of promises we want someone or something to make to us and make possible for us. This cluster of promises could seem embedded in a person, a thing, and institution, a text, a norm, a bunch of cells, smells, a good idea – whatever" (Berlant 2011, 23). This yearning is not just individual but communal in that "Black parents, grandmas, papas, aunties, congregations, Masonic lodges, sorority chapters, and fraternal orders have echoed the determination to ensure that the children in their communities make it to college to earn a coveted degree" (Harris-Perry 2018, n.p.).

The politics of yearning gets complicated though, for as with most things that can be used for good, it can be used against those who try to activate it for a better life. To read the university is to uncover how the university politicizes our yearnings for its own capitalist futurity and in so doing co-opts and distorts them while disavowing the institution's reliance on our longings. Using the example of the movements of the 1960s and 1970s, we witness how a yearning for justice, also motivated by Black, Brown and women students and scholars, particularly with respect to mobilizing education as a tool to achieve social justice, and how the university has tried "to capture, co-opt, and manage their struggles" (Meyerhoff 2019, 162), all the while congratulating itself on the great job higher education has done for women, and Black and Brown folks simply by giving them an opportunity to access the space of reason. I remember at one institution where I formerly worked, working-class students were asked by upper administrators to write letters of gratitude to wealthy white alumni. Those alumni had been given a list of Black and Brown working-class students they could "sponsor," and without knowing or caring to know anything about these students, they sent money to the university to sponsor for themselves a student in need. I received this information from a student who was refusing to write such a letter to an anonymous donor because, in their words, it felt "cringy," but meanwhile the university administrator had emailed that if the student, a Latino from the South Side of Chicago, did not show their gratitude to this wealthy white person, their sponsorship would be withdrawn and given to another student.

This is what Ahmed means when she writes of "conditional hospitality." As mentioned earlier in this chapter, Black and Brown students provide the university with the diversity it seeks, but through the politics of indifference they are assimilated as props to the neoliberal capitalist institutions, who can then use their inclusion against them. The university responds with hostility and anger when they try to speak up and out, or according to hooks, "try to disengage from obligations, communicating the need for care for themselves and their health" (hooks 2005, 67). "Look, see how much we have given you, what we have done for you, don't be ungrateful, do what we ask and don't complain!" The most disturbing part about this conditional type of inclusion is the development of a control rhetoric which takes advantage of yearning and uses it against the marginalized and minoritized to absolve the university from the systems of injustice it continues to perpetuate: through employing a politics of indifference, through including us, but not providing adequate resources, and then blaming us for perceived failures. If we do not complete a course of study or graduate, or if we drop out, it determines that we "didn't want it enough," so that in fact our *lack* of desire must be the reason for failure. It is this rhetoric that tries to hide the institution's "complicity in constructing the boundaries of individualized students themselves, as through teachers instilling fear, anxiety, shame, and pride in students by subjecting them to exams and grades" (Meyerhoff 2019, 95). Exams and grades are an accepted metric for evaluation, but maintaining that these quantitative measurements set the only bar for proving worthiness and desire to succeed is problematic as they do not take into consideration larger societal inequities and exam bias that filter into classrooms. We look forward and yearn so much toward making it, "of being able to *be somewhere* and to make life, exercising existence as a fact, not a project. In other words, in this version of transnational class fantasy, mobility is a dream and a nightmare. The end of mobility as a fantasy of endless upwardness, and the shift to the aspiration toward achieving an impasse and stop-loss, is a subtle redirection of the fantasy bribes transacted to effect the reproduction of life under the present economic conditions" (Berlant 2011, 179).

This use and misuse of the affective for its own gain, engaging in what Meyerhoff refers to as an "emotional economy" works to continue (re)making "and stabiliz[ing] the boundaries and surfaces of entities in the liberal-capitalist imaginary. Believing that you have to choose between these two imagined potential paths of "graduate" or "dropout" – and imagining potential pride or shame – is co-constitutive with viewing oneself as an autonomous, bounded, responsible individual" (2019, 94) within this capitalist world. Quoting hooks again, advanced capitalism affects "our capacity to see, that consumerism ...[has] take[n] the place of that predicament of heart that called us to yearn for beauty. Now many of us are only yearning for things" (hooks 1990, 104) which we as autonomous individuals must work to achieve, and which we believe can be had via higher education; because who truly likes a handout? This of course is a rhetorical question, but what we are left with is mere access to the spaces of education to propel the will of capitalism, and not for our own good. And so the university uses human subjects as measurements – how many of the included dropped out, what percentage graduated? – and uses these measurements to appeal to the yearnings of others, still holding out the institution as good/exceptional at diversity. hooks expresses it best when she describes the two houses in which she grew up:

> Old folks shared their sense that we had come out of slavery into this free space and we had to create a world that would renew the spirit, that would make it life-giving. In that house there was a sense of history. In the other house, the one I lived in, aesthetics had no place. There the lessons were never about art or beauty, but always only to possess

things. My thinking about aesthetics has been informed by the recognition of these houses: one which cultivated and celebrated an aesthetic of existence, rooted in the idea that no degree of material lack could keep one from learning how to look at the world with a critical eye, how to recognize beauty, or how to use it as a force to enhance inner wellbeing; the other which denied the power of abstract aestheticism... I could see in our daily life the way consumer capitalism ravaged the black poor, nurtured in us a longing for things that often subsumed our ability to recognize aesthetic worth or value (1990, 104).

To experience the university in this particular affective way (which I admit not all women and Black and Brown folks do, but enough of us do to make this point important) is to imagine it as the vehicle for a better future – to desire that future and by extension attend the university, to fulfill that desire of entering that space and experiencing on the one hand the feeling that comes with that, the joy, the pride, the exuberance, while simultaneously experiencing the thrill of learning, of building community and graduating. But on the other hand, the frustrations, anxieties and disappointments that turn up when we apprehend the injustice of the space is emotionally taxing to say the least. Economic stresses compound this anxiety: the stress of working to afford school and all the living expenses that go with it are added to the anxiety about achieving a lucrative career to make such a large financial and time investment worthwhile may lead to the students' disillusioning experience of "pressure ... to curtail their imagination during college years ...[and] [f]ind a major that guarantees a good job ... doing as many internships as possible to grease your way into the narrowest doorways that lead to corporate success" (Chatterjee and Maira 2014, 332). In this way dreams give way to the commands of capitalism and fear of debt as well as the compounded – gendered and racialized – fear of being "proven" unworthy to achieve a multigenerational yearning, a yearning which has itself been capitalized upon by the university at considerable expense (financial and emotional).

Three-quarters of students work just to get by, and the time that even part-time work requires unequally impacts their ability to study and, in turn adversely impacts their grades. The success gap is to no small degree a matter of economic privilege. According to a 2005 report of the Advisory Committee on Student Financial Assistance, 29 percent of working-class students, the majority of whom are Black and Brown, "work more than thirty-five hours a week, and of them a majority (53 percent) fail to graduate" (Prashad 2014, 332). For those students who leave without graduating, the feelings of failure and valuelessness wrapped up in that feeling can be devastating. And even for the ones who graduate, feelings also accompany the inability to repay loans after graduation (Ficklen and Stone 2002, 11–13) and feeling of inadequacy if one cannot get a job.

And so we see how the yearning is impacted by the lack of "freedom of the student to enjoy the world of ideas ... ideological and intellectual freedom necessary for critical thought and expression ... to think outside of their indebtedness" (Prashad 2014, 332). Yearning to be let in also means yearning eventually to be let out with the same sense of optimism as when one arrived. The university is a spatiotemporal construct of 4- to 5-year cycles, and as such, the yearning to be "in" also means that one is thinking ahead to what will come when one leaves. Within this spatiotemporal construction, the politics of indifference is sandwiched between what appears to be an intentionality of letting in and letting out. But release, which has to be honorable, bears hidden challenges for racialized/gendered students as well, which are potentialized by the politics of indifference. As such, what Berlant calls these "[p]recarious bodies ... are not merely demonstrating a shift in the social contract, but in ordinary affective states. This instability requires, ... embarking on an intensified and stressed out learning curve about how to maintain footing, bearing, a way of being, and new modes of composure amid unraveling institutions and social relations of reciprocity" (2011, 197). Berlant continues, "Any object of optimism promises to guarantee the endurance of something, the survival of something, the flourishing of something, and above all the protection of the desire that made this object or scene powerful enough to have magnetized an attachment to it" (2011, 48).

There are also the ways in which one's yearning to be included and to belong also pits one against others who want the same thing. For example, Imani Perry writes of how the desire for inclusion by women in the social contract in the nineteenth century "took place largely on the terms of patriarchs and ladies, calling for expansions to the numbers of people included in those categories, over and against those classified as 'nonpersons'" (2018, 41). This example is used to show how when inclusion is predicated on being folded into a system that has historically been oppressive to those seeking inclusion, it can produce a covert culture of competition between/against those most alike, particularly when scarcity is baked into the currency and social capital is doled out by the powerful and comfortable.

As I mentioned above, these are not universal and homogenous experiences for Black and Brown people and women navigating the university. Nirmal Puwar points out that, for example, those who come from "an elite background will have a habitus that is much more in keeping with the demands of the field than those who have not been immersed in this environment. This will occur even while they may 'feel the weight' of the whiteness of organisations and, in this respect, will have occasions where they feel like 'fish out of water', while whiteness is invisible to others, male and female" (2004, 127). These words put me in mind of how folks feel the university on a spectrum. Anneeth Kaur Hundle writes that the university can be experienced by some as "both 'homes and non-homes'" (2019, 303), "sites of both estrangements and intimacies" (318). Inhabiting the space can cause "[d]iscomfort, pain, and tension ... Yet the University can also be a home for marginalized beings and bodies: a sanctuary that allows for intellectual liberation, recovery and healing, self and subject making [as COVID-19 revealed many marginalized and poor students had no home to return to when universities closed]. It engenders new forms of community and family for dispossessed intellectuals. We push the limits of the borders, walls, and ceilings that we are confronted with as we remake and fashion universities as homes – often needing to travel to new havens in the process, becoming itinerant intellectuals" (Hundle 2019, 303).

With that in mind however, based on the research into discrimination within the university, it is safe to assume that many people do share these confounding and tortuous experiences. "The intensity of the need to *feel* normal [not as the non-human or non-person] is created by economic conditions of nonreciprocity that are mimetically reproduced in households that try to maintain the affective forms of middle-class exchange while having an entirely different context of anxiety and economy to manage" (Berlant 2011, 180). Just as I appreciate my mother's and grandmothers' encouragement to reach for the stars, I understood how absolute was their familial and community pride, and lived for the day I could lend financial support to my working poor family, the simultaneous pressure to perform, to be and feel normal through achieving middle-class success, was (and still is) often palpable as I push/ed myself in unimaginable ways. hooks writes about the ways in which "black people, and black women in particular, are so well socialized to push ourselves past healthy limits" (2005, 40–41) and that "[p]ractically every black woman [she] know[s] spends way too much of her life-energy worried and stressed about money. Since many of us are coming from economic backgrounds where there was never enough money to make ends meet, where there was always anxiety about finances" (hooks 2005, 43). Many of us do this living away from the same family, community, and friends who we hope that our hard work would eventually benefit and "we may need to evaluate whether or not we are gaining in overall quality of life by being ...[there] rather than where our love and support is" (hooks 2005, 44). These are not just cerebral decisions of this versus that. These dilemmas are deeply felt, and we negotiate our feelings in the vein of not wanting to disappoint our folks, or that we have made it this far and so we cannot give up now, especially since so many people sacrificed for us to be here. Black people according to hooks, "do not have this sense of 'entitlement' [unlike Nance, the parent mentioned in the Introduction who wrote the Op Ed complaint about pronouns and performances of land acknowledgements]. We are not raised to believe that living well is our birthright" (2005, 45) so that when the opportunity comes, the responsibility is great.

The pressures of existing in minoritized and marginalized bodies then means that we cannot simply experience the university in a purely cerebral way, as "the life of the mind." To imagine we may simply choose to do so is to (dis)miss the complex ways in which we inhabit the university and to miss the ways in which the politics of indifference operates; so that we feel the university with a complex mix of joy, pride, happiness, "confusion, disorientation, anxiety, and apathy, mixed with concerns about our future relations with what we need and care about in life – our employment, health, family, housing, food, and so forth" (Meyerhoff 2019, 23), all wrapped up in the institution. This recognition of how systems of domination make us feel, according to Nash, is the gift that affect theory has given us (2019, 30), thus allowing us get to "the praxis of politics – the whole dynamic by which latent human needs are expressed in political terms and, by being formulated, become the conscious demands of a section of the society, around whom a political agitation can be built, maintained, and carried. Of course, this is the radical model of politics" (Hall 2017, 88). A radical model is what I am suggesting in this book by naming both the politics of indifference and the politics of yearning, and reading the university.

We inhabit the university as bodies, walking around it doing what we think are ordinary things which normal people who inhabit that space do. But at the same time, the university inhabits us. With students in particular Black and Brown students and women, there is also the added pressure not to show how one is affected by the university, in effect kill one's affective so as not to appear in the number of stereotypical ways one's body shows up for showing one's authentic feelings – take for example the angry black women troupe. Sara Ahmed writes about this in her text *The promise of happiness:* "the body of color is attributed as the cause of becoming tense, which is also the loss of shared atmosphere (or we could say that sharing the experience of loss is how the atmosphere is shared) …you do not even have to say anything to cause tension. The mere proximity of some bodies involves affective conversion" (2010, 67). We become as indifferent to our pain as others, just get through it, grin and bear, persevere because we are strong, but "[b]eing 'used to pain' does not mean that we will know how to process it so that we are not overwhelmed or destroyed by grief" (hooks 2005, 78). When we acknowledge this grief, it can become a catalyst for political action.

The three, politics of indifference, apathetic methodology of power, and politics of yearning, all work together to operationalize and justify exclusion through the uses of measurement of the bodies which they include/exclude. These three operate together to allow the university moves that are grounded in "facades of inclusion without any change to dominant and unequal power structures or knowledge bases, and this is often the crux of the matter" (Sultana 2019, 35). There is no attempt to truly understand the person outside of the utility for the academy that their body provides. As Glissant writes, "If we look at the process of 'understanding' beings and ideas as it operates in Western society, we find that it [is] founded on an insistence of this kind of transparency. In order to understand and therefore accept you [,] I must reduce your density to this scale of conceptual measurement which gives me a basis for comparisons and perhaps for judgements" (1997, 204).

To be human is to feel profoundly (Quashie 2012, 72). There is a lot for us to feel about being "included" in the university, and today as we deal with the university's lack of response to the pandemic (and other crises disproportionately affecting us), we cannot attempt a "deadening ... [of our] affective tensions" (Meyerhoff 2019, 46) toward the university in order to survive, by investing in ideals of a future of imaginative security which we are taught and which we believe comes from an investment in education. According to Alexander, because we have been so subjected to scientific, academic dismemberment, the work of decolonization has to include our deep yearning "that is both material and existential, both psychic and physical, and which, when satisfied, can subvert and ultimately displace the pain of dismemberment ... This yearning to belong is not to be confined only to membership of citizenship in community, political movement, nation, group, or belonging to a family, however constituted, although important ...we recognized this yearning as a desire to reproduce home in 'coalitions'" (Alexander 2005, 281–282). Kevin Quashie refers to this recognition of our deep interior emotionality as the sovereignty of quiet.

While the university coopts the yearnings of Black and Brown people and women for its own use, Quashie suggests, much like Alexander, that there is an interior, "raucous and full of expression ... [that is] the source of human action – that anything we do is shaped by the range of desires and capacities of our inner life" (2012, 4). Borrowing from Alexander and Quashie, I posit that we must attend to these desires in ways that validate them in order to better understand our feelings of anger and frustration around how we are included into the university and carry the full weight of our identity groups on our shoulders, how "affective transactions ... take place alongside the more instrumental ones" (Berlant 2011, 167) such that actual university actions – budget cuts, tuition hikes, lack of resources and more – can be read and felt in specific ways by certain bodies. It helps us to understand that when we are told "to look 'spry and smart'; this conspiracy of appearances acts to repudiate the claims of pain" (Hartman 1997, 39).

We validate the ways in which we feel this transaction by pausing in order to try to understand "the impasse of living in the overwhelmingly present moment" (Berlant 2011, 49). In the next chapter, I pause to more deeply examine university inclusion and diversity initiatives today as technologies of the politics of indifference, using the stories of Black and Brown students and women and in particular the personal narrative, or what Perry refers to as "mini-narrative" of one Black woman student whose own thoughts and feelings about how she feels the university "serve[s] as models of how we read layers of domination at work" (2018, 12) to bring home how the politics of indifference works to politicize our yearnings. I end this chapter with a quote from Audre Lorde:

> "But sometimes we drug ourselves with the dream of new ideas ... There are no new ideas, only new ways of making them felt, of making them real. For within these structures which we live beneath, defined by

profit, by flat linear power, by institutional dehumanization, our feelings were not meant to survive ... Our feelings were meant to kneel to thought as women were meant to kneel to men. But women have survived, and our feelings have survived. ... They lie in our dreams, they lie in our poems, and it is in our dreams and our poems that point the way to our freedom. ... Learn to love the power of your feelings, and to use that power for your good" (2009, 122).

Reading the university brings in shad(e)y theoretics to pick apart what's ugly and helps tell a hidden truth – it allows the complexity of the affective relationship with higher education to come out. For those who are hurt by and disappointed with how the university does not fit or feel right, for whom it has not provided the pleasures it promised despite all the investments made, who see paths opened up but also seeming to simultaneously close or narrow, reading causes us to question our past yearnings, to engage in the journey toward being more conscious of why, through making affectual connections. Understanding the ways in which we feel the university (and how we are assessed by it) leads our investigation toward a sense that this feeling is about a deeper systemic issue. This reading practice, as Nash tells us in her theory of affect, helps us to get to the bottom of the "...felt life of racial gendered violence, and a critical analysis of the myriad spaces where this violence unfolds (Nash 2019, 123)"

Note

1. These protection might admittedly not be extended depending on the race and gender of the person, and the political climate at the time of the political activism as seen from the opposition to WWI in the example.

References

Ahmed, Sara. *Strange Encounters: Embodies Others in Post-Coloniality.* London: Routledge, 2000.

Ahmed, Sara. *The Promise of Happiness.* Durham: Duke University Press, 2010.

Ahmed, Sara. *On Being Included: Racism and Diversity in Institutional Life.* Durham: Duke University Press, 2012.

Alexander, M. Jacqui. *Pedagogies of Crossing: Meditation on Feminism, Sexual Politics, Memory, and the Sacred.* Durham: Duke University Press, 2005.

Andrews, Kehinde. "The Black Studies Movement in Britain: Becoming an Institution, Not Institutionalized." In *Dismantling Race in Higher Education: Racism, Whiteness and Decolonizing the Academy*, edited by Jason Arday and Heidi Safia Mirza, 271–287. London: Palgrave McMillian, 2018.

Ang, Ien. "I'm a Feminist but ... 'Other' Women and Postnational Feminism." In *Oxford Readings in Feminism: Feminism & 'Race,'* edited by Kum-Kum Bhavani, 394–409. Oxford: Oxford University Press, 2001.

Ansfield, Bench. "Still Submerged: The Uninhabitability of Urban Redevelopment." In *Sylvia Wynter: On Being Human as Praxis*, edited by Katherine McKittrick, 124–141. Durham: Duke University Press, 2015.

Berlant, Lauren. *Cruel Optimism*. Durham: Duke University Press, 2011.

Berry, Daina R. *The Price for Their Pound of Flesh: The Value of the Enslaved, from Womb to Grave, in the Building of a Nation*. Boston: Beacon Press, 2017.

Bourne, Randolph S. "The Idea of a University." In *War and the Intellectuals: Collected Essays, 1915–1919*, edited by Carl Resek, 152–153. Indianapolis: Hackett Publishing Company, 1999.

Byrd, Rudolph P., Johnetta Betsch Cole and Beverly Guy-Sheftall. *I Am Your Sister: Collected and Unpublished Writings of Audre Lorde*. New York: Oxford University Press, 2009.

Chatterjee, Piya and Sunaina Maira. "Introduction: The Imperial University." In *The Imperial University: Academic Repression and Scholarly Dissent*, edited by Piya Chatterjee and Sunaina Maira, 1–50. Minneapolis: University of Minnesota Press, 2014.

Chinn, Sarah E. *Technology and the Logic of American Racism: A Cultural History of the Body as Evidence*. London: Continuum, 2000.

da Silva, Denise Ferreira. *Toward A Global Idea of Race*. Minneapolis: University of Minnesota Press, 2007.

da Silva, Denise Ferreira. "1 (Life) ÷ 0 (blackness) = ∞ - ∞ or ∞ /∞: On Matter Beyond the Equation of Value." *e-flux journal* no. 79 (2017): 1–11.

Davies, Carole Boyce. *Black Women, Writing and Identity: Migrations of the Subject*. New York: Routledge, 1994.

Davies, Carole Boyce. "From Masquerade to *Maskarade*: Caribbean Cultural Resistance and the Rehumanizing Project." In *Sylvia Wynter: On Being Human as Praxis*, edited by Katherine McKittrick, 183–202. Durham: Duke University Press, 2015.

Davis, Angela Y. *Women, Race, and Class*. New York: Vintage Books, 1983.

Davison, Sally, David Featherstone and Bill. Schwarz. "Introduction: Redefining the Political." In *Political Writings: The Great Moving Right Show and Other Essays*, edited by Sally Davison, David Featherstone, Michael Rustin and Bill Schwarz, 1–15. Durham: Duke University Press, 2017.

Ficklen, Ellen and Stone Jeneva. *Empty Promises: The Myth of College Access in America: A Report of the Advisory Committee on Student Financial Assistance*. Washington DC: Advisory Committee on Student Financial Assistance, 2002.

Glissant, Édouard. *Poetics of Relation*. Ann Arbor: University of Michigan Press, 1997.

Grossberg, Lawrence. "History, Politics and Postmodernism: Stuart Hall and Cultural Studies." *Journal of Communication Inquiry* Vol. 10, no. 2 (1986): 61–77. doi: 10.1177/019685998601000205.

Guillory, John. "[1996] Preprofessionalism: What Graduate Students Want." *Profession* (2012): 169–178.

Hall, Stuart. "And Not a Shot Was Fired [written in 1991]." In *Political Writings: The Great Moving Right Show and Other Essays*, edited by Sally Davison, David Featherstone, Michael Rustin and Bill Schwarz, 266–274. Durham: Duke University Press, 2017.

Hall, Stuart. "Political Commitment [written in 1966]." In *Political Writings: The Great Moving Right Show and Other Essays*, edited by Sally Davison,

David Featherstone, Michael Rustin and Bill Schwarz, 85–106. Durham: Duke University Press, 2017.

Harris, Adam. "The Death of an Adjunct." *The Atlantic* (2019).

Harris, Cheryl I. "Whiteness and Property." *Harvard Law Review* Vol. 106, no. 8 (1993): 1707–1791.

Harris-Perry, Melissa. "What It's Like to Be Black on Campus Now." *The Nation* (2018). https://www.thenation.com/article/archive/what-its-like-to-be-black-on-campus-now/.

Hartman, Saidiya V. *Scenes of Subjection: Terror, Slavery, and Self-Making in Nineteenth-Century America*. New York: Oxford University Press, 1997.

Hong, Grace Kyungwon. "'The Future of Our Worlds': Black Feminism and the Politics of Knowledge in the University Under Globalization." *Meridians* Vol. 8, no. 2 (2008): 95–115.

hooks, bell. *Yearning: Race, Gender, and Cultural Politics*. Boston: South End Press, 1990.

hooks, bell. *Sisters of the Yam: Black Women and Self-Recovery*. Cambridge: South End Press, 2005.

Hundle, Anneeth Kaur. "Decolonizing Diversity: The Transnational Politics of Minority Racial Difference." *Public Culture* Vol. 31, no. 2 (2019): 289–322. doi: 10.1215/08992363-7286837.

Lauter, Paul. "From Adelphi to Enron." *Academe* Vol. 88, no. 6 (2002): 28–32. doi: 10.2307/40252437.

Lomax, Tamura. *Jezebel Unhinged: Loosing the Black Female Body in Religion and Culture*. Durham: Duke University Press, 2018.

McKittrick, Katherine. "Plantation Futures." *Small Axe* Vol. 17, no. 3 (2013): 1–15. https://read.dukeupress.edu/small-axe/article-abstract/17/3%20(42)/1/33296/Plantation-Futures.

McKittrick, Katherine. "Axis, Bold as Love: On Sylvia Wynter, Jimi Hendrix, and the Promise of Science." In *Sylvia Wynter: On Being Human as Praxis*, edited by Katherine McKittrick, 142–163. Durham: Duke University Press, 2015.

McKittrick, Katherine. "Rebellion/Invention/Groove." *Small Axe* Vol. 20, no. 1 (2016): 79–91. doi: 10.1215/07990537-3481558.

Mendoza, Breny. "Coloniality of Gender and Power: From Postcoloniality to Decoloniality." In *The Oxford Handbook of Feminist Theory*, edited by Lisa Disch and Mary Hawkesworth, 2–26, 112. Oxford: Oxford University Press, 2012.

Meyerhoff, Eli. *Beyond Education: Radical Studying for Another World*. Minneapolis: University of Minnesota, 2019.

Mignolo, Walter D. "Sylvia Wynter: What Does It Mean to Be Human." In *Sylvia Wynter: On Being Human as Praxis*, edited by Katherine McKittrick, 106–123. Durham: Duke University Press, 2015.

Moffett-Bateau, Courtney. "American University Consensus and the Imaginative Power of Fiction." *Journal of Critical Ethnic Studies Association* Vol. 4, no. 1 (2018): 84–106.

Nash, Jennifer C. *Black Feminism Reimagined: After Intersectionality*. North Carolina: Duke University Press, 2019.

Perry, Imani. *Vexy Thing: On Gender and Liberation*. Durham: Duke University Press, 2018.

Prashad, Vijay. "13 Teaching by Candlelight." In *The Imperial University: Academic Repression and Scholarly Dissent*, edited by Piya Chatterjee and Sunaina Maira, 329–342. Minneapolis: University of Minnesota Press, 2014.

Puwar, Nirmal. *Space Invaders: Race, Gender and Bodies Out of Place.* Oxford: Berg, 2004.

Quashie, Kevin. *The Sovereignty of Quiet: Beyond Resistance in Black Culture.* New Jersey: Rutgers University Press, 2012.

Robinson, Cedric J. *Black Marxism: The Making of the Black Radical Tradition.* Chapel Hill: The University of North Carolina Press, 2000.

Said, Edward W. "Identity, Authority, and Freedom: The Potentate and the Traveler." *boundary 2* Vol. 21, no. 3 (1994): 1–18.

Sharpe, Christina. "Black Studies: In the Wake." *The Black Scholar* Vol. 44, no. 2 (2014): 59–69. doi: 10.1080/00064246.2014.11413688.

Slaughter, Sheila and Gary Rhoades. *Academic Capitalism and the New Economy: Markets, State, and Higher Education.* Baltimore: John Hopkins University Press, 2004.

Spillers, Hortense J. "Mama's Baby, Papa's Maybe: An American Grammar Book." *Diacritics* Vol. 17, no. 2 (1987): 65–81.

Stein, Sharon. "Higher Education and the I'm/possibility of Transformative Justice." *Journal of Critical Ethnic Studies Association* Vol. 4, no. 1 (2018): 130–153.

Sultana, Farhana. "Decolonizing Development Education and the Pursuit of Social Justice." *Human Geography* Vol. 12, no. 3 (2019): 31–46. doi: 10.1177/194277861901200305.

Taylor, Kate. "College Admissions Scandal." *The New York Times* (2020). https://www.nytimes.com/news-event/college-admissions-scandal.

Wilder, Craig Steven. *Ebony and Ivy: Race, Slavery, and the Troubled History of America's Universities.* New York: Bloomsbury Press, 2013.

Young, Iris Marion. *Justice and the Politics of Difference.* Princeton: Princeton University Press, 1990.

Young, Iris Marion. *Intersecting Voices: Dilemmas of Gender, Political Philosophy, and Policy.* Princeton: Princeton University Press, 1997.

Young, Iris Marion. "Structural Injustice and the Politics of Difference." AHRC Centre for Law, Gender, and Sexuality (2005).

3 Feeling inclusion/exclusion

diversity operates in apolitical and often antipolitical ways to selectively usher a few bodies into exclusive institutions.

– Nash 2019, 24

I have a hard time accepting diversity as a synonym for justice. Diversity is a corporate strategy. It's a strategy designed to ensure that the institution functions in the same way that it functioned before, except now you have some black faces and brown faces. It's a difference that doesn't make a difference.

Diversity without structural transformation simply brings those who were previously excluded into a system as racist, as misogynistic, as it was before.

– Davis, 2015

Of course, it is all data: Every bit of you is being collected – your accounts, your e-mails, your demographics ... The data are "profitable" and traded on.

– Perry 2018, 137

As detailed in the previous chapters, the enlightenment knowledge project that is the university was founded on the making of the normative modern subject based on a racist, sexist, imperialist understanding of this subject as white, cisgender and male. From its inception through to today, this knowledge project continues to proliferate within the university often engaging violent and exclusionary practices, resulting in violation, displacement and dispossession (Crawley 2018, 10). As times have changed however, it has become harder for the university to engage in the overtly racist science mentioned earlier in this text, and also to exclude people solely on the basis of race and gender. This, however does not mean that the university no longer engages in it, rather that the type of racist, sexist, capitalist violence of the past which resulted in exclusion "cannot be easily folded into the logics of inclusion, diversity, and multiculturalism" we see today (Crawley 2019, 11), and as such this violence also becomes harder

DOI: 10.4324/9781003019442-4

to detect. In this current period, struggles for inclusion have been exacerbated by a shift away from legal affirmative action, which was supposed to redress the historical inequities mentioned above, as well as the pivot to "diversity initiatives" brought on by the complexities of the 2003 *Grutter v. Bollinger* Supreme Court decision. In this climate, the university is not required to address issues of social inequities within its ranks since the move toward diversity focused less on the group and institutional accountability and more on individualist inclusion as a solution to historical mass injustice (Hundle 2019, 292).

In this chapter, I argue that diversity and inclusion are derived from the same systems of classification and categorization practiced and enforced by the neoliberal university, and as such they continue the historical legacy of discipline and symbolic violence perpetuated by the university (Alexander 2005, 141). Despite decades of internal and external activism calling for equal opportunity in education, the university, even as it makes a claim of being a liberal institution, continues to align with the state, and remains critical in the contemporary fight around normalized culture, by creating, supporting, and contesting knowledge upon which state policies are based (Chatterjee 2014, 7). In the 1960s and 1970s, liberal academic ideologues called for changes including greater representation of Black and Brown students and women and changes to the curriculum. For example, Bradley writes of that time, "[a]s a result of rallies we got courses in 'black literature' and 'black history' and a special black adviser for black students and a black cultural center … rotting white washed house on neither edge of campus … reachable … by way of a scramble up a muddy bank … And all those new courses did was exempt the departments from the unsettling necessity of altering existing ones" (1982 cited in Wynter 2015, 20). What Bradley describes is analogous to "separate but equal" by a different name, as students had to contend with literal physical obstacles as well as apathy by the administration toward diversity and inclusion initiatives at the time, turning the victory into what felt like punishment. While it is true that the 1960s and 1970s were an unprecedented time where the university flung open its doors to welcome students previously unwelcomed, as Ahmed asserts, this welcome positioned them as "not yet a part, a guest or stranger, the one who is dependent on being welcomed, the one whose arrival is conditional on the will of those who are already here" (2018, 333–334).

In this chapter, I discuss how we must not mistake access to education for true progress toward equity and educational justice. As Christi Smith points out, "[t]hroughout the history of higher education, access has not secured inclusion" (2016, 143) in its broadest sense. I also demonstrate herein the specific ways in which this access, wrapped up in a seemingly paradoxical inclusive exclusion, is felt by those to whom it is extended. Take, for example, the inclusion of women with the introduction of coeducation colleges in the 1830s and 1840s. Feminist and abolitionist activism closely matched a steady growth in the need for educators and nurses during and after the Civil War (Smith 2016, 145), and as a result, more women were given access to education to prepare them for their socially predestined roles as mothers, missionaries, nurses and teachers (Smith 2016, 142–143). According to Smith, while women at the time demanded inclusion and access to private universities they, specifically white bourgeois women, were granted access to education mainly through the creation of separate women's institutions, such as Vassar and Smith Colleges, "expressly designed for the 'unique biological needs' of women" (2016, 144) or annexes at institutions like Harvard and Columbia where women were "permitted to take exams and gain limited recognition for intellectual accomplishment, but were barred from regular academic life. For instance, Barnard and Radcliffe Colleges grew out of annex programs" (Smith 2016, 144). These women were strictly monitored and measured under the guise of medical care, as critics used biology to argue that women's bodies were both physically and mentally inferior, and that their mental exertion would be dangerous to their health (Smith 2016, 147–148). Reproductive fitness, in particular, was a concern – a holdover from the pre-modern concept of the "wandering womb," and the nineteenth-century diagnosis of "hysteria" as a sign of women's inherent instability.

Those advocating for women's inclusion at the time appealed to fairness (Smith 2016, 152), a rationale still used today to advocate for access for other marginalized groups. However, the optics and public practices of "fairness" have still largely brought mixed results, and still attract pushback not only from academics but also from lay people like Nance, as described in the Introduction to this book. Under the guise of protecting them, women were included in higher learning in very restricted ways, so as to prevent them from developing "a distaste for the pleasures of domestic life [which] would disqualify them for their duties in both family and society" or threaten their reproductive capacity (Smith 2016, 154–155). To address this concern, the university attempted to measure the impact of education upon women's bodies. What the university was in fact doing was "protect[ing] firm status boundaries between blacks and whites, men and women" (Smith 2016, 151–152), because as examined in the previous chapters, gender and race are always imbricated in the matrices of oppression.

Women's bodies were considered strange and unnatural to the public sphere, and public women – working-class and Black women in particular – were reviled. According to Young, bourgeois gender ideology in the nineteenth century designated the public sphere of commerce and politics as the sole province of men, and women were relegated to the domestic sphere, for they were constructed as incapable of acquiring discipline and self-control inherent in respectable men (1990, 137). According to Smith:

> As women made gains in higher education, their increasing presence was perceived by some as a threat to the pervasive, exclusionary masculine control of a key social institution for categorizing privilege and

status. When elites were pressured to vary established admissions practices, a new campaign was launched to rearticulate the rules governing women's access to symbolic and material resources for upward status mobility: sharp class distinction, biological difference, and racial competition. Just as rising "scientific racism" articulated a theory of polygenism, or the belief that different races developed on separate tracks, academics applied a similar logic to women. Women would be harmed by forms of higher education not designed with their particular, unique biological development at its core. Race, class, and gender are understood as sustained patterns of interlocking inequality, a matrix of domination, deeply embedded in the most personal aspects of both individual and social life (2016, 158).

To prove whether women's bodies were fit for a program of education, scientists used physical appearance as a means of measurement, gauging mental and moral fitness through the perfection of the physical form. They engaged in what Megan H. Glick refers to as "'ocular anthropomorphism' ... or a process of humanization deeply tied not only to the rhetoric of visual representation but also, and primarily, to the rhetorics of sight, ocular consciousness, and the scientific gaze" (2018, 59), in this case photography. According to Glick, scientists like American eugenicist and primatologist Robert Yerkes, for example, used photography to "confirm" and "create the terms of species difference" (2018, 67). Along with other types of physical measurement, scientists also used photography to determine human fitness by measuring posture, which was of course connected to other eugenicist premises about the perfect human body and brain being white and male, and which according to them could be proved through these measurements. This form of assessment was applied to women seeking higher learning. A practice widely adopted in the late 1800s of photographing and assessing women's posture, and then forcing them to take classes to fix said posture in order to graduate, was still in use in the 1960s.

  The practice of so-called posture portraits developed from the idea that proper posture was under increasing threat, and could result in physical degeneration. In essence, the argument went, an erect stature was necessary to prevent internal organs from being crushed,[1] and bad posture was linked to bad habits and poor character, and could impede the breeding abilities of white women (Vertinsky 2007, 297-298). There were many debates at the time between physicians regarding the composition of bourgeois, white women's bodies; however, as Strings notes, physicians "could agree that the dietary habits and body size of these women had to be monitored. The reproductive capacity of these women was threatened due to their unwise choices, and the nation would suffer accordingly" (2019, 194). According to Vertinsky, seeing physique and character as linked to posture was in fact a way of "substantiating a priori beliefs about class, race, and gender and highlighting the growing ambiguity inherent in the term normal" (2007, 297).

Posture she writes, "was applied easily to racial analysis, with the argument that those of European descent had erect spines and straight bones allowing graceful deportment to further the contrast to the stooped posture and flat feet of non-white, less civilized races. Ivy leaguers were exhorted to demonstrate their erect (upper class) carriage and linearity through good posture" (2007, 300). The scientific slippage linking the inclusion of white women to higher education to possible deformity, degeneration of habits and character to bad posture, and bad posture to people of non-European descent and anxieties surrounding gender and class, justified the measuring of women's bodies, and the seemingly objective exclusion of Black and Brown people in general, people from the working classes, and those with physical disabilities, from higher education. This train of logic fits in lockstep with general attitudes toward ability. According to Mia Mingus, the concept of ability and ableism in general is crosscutting, in that

> ableism dictates how bodies should function against a mythical norm – an able-bodied standard of white supremacy, heterosexism, economic exploitation, moral/religious beliefs, age and ability. Ableism set the stage for queer and trans people to be institutionalized as mentally disabled; for communities of color to be understood as less capable, smart and intelligent, therefore "naturally" fit for slave labor; for women's bodies to be used to produce children, when, where and how men needed them; for people with disabilities to be seen as "disposable" in a capitalist and exploitative culture because we are not seen as "productive" (2011, n.p.).

Thus, the academy was a stronghold of pseudoscientific opposition to the inclusion of non-white and non-male bodies.

In 1873, Edward H. Clarke, a Harvard medical professor, stated in a written appeal against coeducation as a threat to a healthy body, "that women's reproductive system 'organized' the development of body and mind, which 'lends to her development and to all her work as rhythmical or periodical order, which must be recognized and obeyed'" (Smith 2016, 159). A few years later, when D.A. Sargent, also at Harvard, began detailing anthropometric measurements to compare and interpret variability in parts of the body, including posture, he was continuing the search for a racist, classist ideal of normality which also discriminated based on gender. These techniques, including photography as a tool to measure bodies, began to proliferate at other institutions of higher education. The commitment to the normal ideal body saw physical educators at schools collaborate with the American Posture League, adopting methods including nude or semi-nude side, front, and rear posture photographs and posture measurements (Vertinsky 2007, 298). These posture portraits were taken at colleges and universities across the United States from the late nineteenth century to the late 1960s, including private women's colleges but also at colleges like Yale and Harvard.

At numerous women's colleges, posture photos were taken of all incoming first-years and pre-graduation seniors.

For example, Vassar College began measuring students and notating their height and girth beginning in 1884 for the basic course in Physical Education or Hygiene and Physical Education. By 1919, Vassar had established posture committees and held posture drives that attempted to inspire students to carry themselves more correctly (Vassar Encyclopedia 2005). As English Professor Elizabeth Daniels described the method at Vassar: "In your freshman year, you went over to the gym... where ... a pro-tem photo booth [was set up]. You changed into an angel-robe before entering the booth, shed the robe temporarily while your nude profile and rear views of your body were recorded, put the angel robe back on, and left" (Vassar Encyclopedia 2005, n.p.). This process also took place at Connecticut College where girls wore "angel robes" before stripping down to be evaluated (Connecticut College News 1928 and 1930), and at Wellesley where it was also recorded whether the student was underweight or overweight (Chen 1933). These medicalized posture portraits were still mandatory up until the late 1960s at some schools, and because they were based in eugenics, and therefore inherently sexist, racist and ableist, their practice meant that Black, Brown and disabled women faced added barriers to inclusion. When we think back to McKittrick's theorizing on the social construction of space, it is clear that while at the time of their founding women's colleges had no explicit admissions policy addressing the issue of race, in many cases they did have quotas, and used *a priori* racist policies to exclude black women, even though in rare instances black women were able to matriculate (Smith 2016, 149–150). By the 1960s, Black women were attending higher learning institutions in record numbers, and yet these racist, sexist, measurement standards still existed. Black and Latina students at Barnard, for example, wrote a poem titled "Good Sisters" in the 1972 year book, The Mortarboard: "There are people, places, and things we'd like to forget:... Those blatantly racist posture pictures, we cannot all have curvature of the spine!!!" (The Mortarboard 1972).

I offer this lengthy example because in 2017 I had the opportunity to interview seven women who had their posture portraits taken in the 1960s. They shared with me how dehumanized they felt being forced to endure having these photographs taken.[2] These women, ranging in age from 78 to 92, even with memories fading in some cases, could recall how they were left in various states of undress alone with the photographer, who was usually male. One of the women who I call Carol, describes the procedure she endured and which varied from institution to institution; there was "a white background and I think there was a line, like a plumb line, a string with a heavy lead weight on the bottom so it would hang straight, and I guess we were supposed to stand in front of that and they probably guided us. And they took the picture to see ... how straight you were." Whatever the variations in the procedure, the way in which these women felt being required to pose nude or semi-nude to have their photos taken and bodies measured can be summed up thusly: it "was one of these things that you grit your teeth and you grin and bear it and it wasn't a choice ... in speaking with friends and talking about it again now ... it gives me this awful feeling- a feeling of panic. I have a feeling that although I don't specifically remember what happened, I remembered the feeling that I had. Otherwise why would I be feeling this now? ... It was a dehumanizing thing ... And actually knowing what I know now I would be furious because I feel that it's a violation" (Angie).

The description above about "grinning and bearing," and the indifference with which student outcry against this routine violation was met, is exemplified in the 1931 comment by Katherine Blunt, then President of Connecticut College, that, "[n]o attempt will be made unduly to influence the girls, but the usual means of furthering health, such as the personal hygiene and nutrition class, medical advice, posture tests, etc. will be continued" (The Day 1931, n.p.). Her statement resonates with how technologies of oppression were utilized to give the impression of complicity and agency of singing and dancing enslaved persons, as described by Hartman and mentioned in the previous chapter, and also with the smiling faces of women and Black and Brown students in university brochures and web pages today. In our contemporary period, "as one admissions officer at an Ivy League institution confessed, part of the rationale behind the trend in requiring photographs as part of college application materials is that schools desire students with 'fit minds and fit bodies'" (Glick 2018, 166). These sexist, racist technologies of oppression remain embedded in inclusion policies and practices today just as in the past, and it can be argued that they are now even more restrictive, for unlike the posture portraits, these photos are required even before orientation to assess and determine who is included.

In fact, I argue that as policies of admission appeal to fairness, it is important to call out how photography is used to play up an idea of fairness that is not reflected in actual practices of equality: note *how* photography is still used today by institutions as a seemingly impartial, documented method of capturing data of and about students' bodies for their use. When, as Ahmed points out, "[t]he promise of diversity is the promise of happiness" (2012, 165), what better way to demonstrate this but through photographs of smiling Black and Brown and women's faces in glossy university brochures? These pictured happy, smiling faces make inclusion appear as the remedy to inequality, such that "[d]iversity work becomes about generating the 'right image' and correcting the wrong one" (Ahmed 2012, 34). Once again, the manipulated conclusions drawn from a photographic image becomes a stand-in for "objective truth" subject to biased interpretation. White supremacist fantasies about the "happiness" of Black and Brown people (especially Black women) within the institution are perpetuated by the use of the medium of photography as measurement.

Ahmed further asserts that while these photographs serve to provide the right image for the university, signaling that the university is open to change and an exceptional beacon for inclusion and diversity, they actually *affect* the bodies, minds, and spirits of those forced to "grin and bear it." Although photographic "evidence," as exemplified in publicity materials, is supposed to signal an institution's success in doing the work of inclusion, and furthermore doing it as a form of care for the newly included, the images actually do the work of exclusion using the ideal images of bodies of those who seek inclusion as evidence against them. Both posture photography and diversity photography signal that the "diverse" bodies have been included based on how they measure up to the university's standards, and thus imply that their continued inclusion remains contingent on meeting and maintaining them. Only now the standards are more insidious. The novel form of the "posture" ideal which the "included" Black, Brown, female students must live up to is an impossible gendered/racialized metric of performed happiness and contentment with things as they are now.

Ahmed's read on the purpose and effect of photographic measurement demonstrates how inclusion projects are part of the "*long durée* accretions of concepts and ideas that continue to be foundational to the neocolonial apparatus of the modern university. By visibilizing [them] … we can begin to ask new questions. For instance, what would it mean to create universities in which marginalized students are not merely tolerated, or assimilated within (new) liberal multiculturalisms and nationalist citizen-subject-making projects, but a focal point for the making of a new university" (Hundle 2019, 301)? In order to create that new world in higher education, our understanding has to come from a place outside of and beyond the metrics. We need to question these metrics and measurements by asking questions like, if the metrics touting "diversity" are true, why hasn't equality prevailed? And how have these metrics turned into yet another statistic utilized by the white supremacist institution to "prove" that diversity and inclusion do not work? In the remainder of this chapter, I demonstrate by way of another example, how we can work to question the ways in which diversity metrics work, visibilizing them, so as to expose their negative impacts on marginalized and minoritized students.

According to Slaughter and Rhoades,

> From 1973 to 1992, the number of students making the direct transition from high-school to college increase from 4 to 7 percent to 62 percent … Although women have made great gains in college enrollment … ethnicity-based inequities persisted. From the 1970s through the 1990s, the numbers and proportion of African-Americans, Hispanics, and Native Americans achieving baccalaureate degrees increased … Yet the wide gap between their college achievement and that of Anglo Americans has continued (2004, 281–282).

As the neoliberal university moves through time, its push for diversity uses difference, according to Jasbir Puar, as a way of producing "new subjects of inquiry that then infinitely multiply exclusion in order to promote inclusion. Difference now precedes and defines identity" (2012, 55). As such, the ways in which these bodies occupy/inhabit the academy as women, and as Black and Brown bodies often requires a posture of survival via legibility. In order to be institutionally legible, those occupying academic borderlands too often "are forced to either translate themselves into the institutional framework of their university ... or remove themselves from their languages that defined their cultural hubs" (Moffett-Bateau 2018, 91-92). These inclusion initiatives, as mentioned above, both discipline and produce complicity, as those newly included seek to find their place in the reorganization of the university to fit current racist neoliberal logics that require "compliance with institutional measures of exclusion, even when it is disguised as diversity and narrated through inclusion" (Moffett-Bateau 2018, 99). According to Anzaldua quoted in Keating, diversity is "treated as a superficial overlay that does not disrupt any comfort zones" (Keating 2009, 205). In other words, the identity of "difference" is admitted, but what that difference means in the context of changing an exclusionary, oppressive institution is a forbidden topic, outside of that "comfort zone." Conditional welcoming is based on keeping the status quo undisturbed and "comfortable."

The inclusion of certain bodies then also requires their management, what Crawley calls "crisis management" (2018, 12), meaning, managing those whose bodies show up as crisis and possessing the potential to create crises. Diversity "operates as a technology of governance that one is disciplined into" (Hundle 2019, 294). This is done through among other things a discourse of scarcity. As a place of rationality and the mind, the university as we saw in the previous chapter is imagined as sacred, admitting only the most prepared mentally and physically for what the rigors of intellectualism will do to them. Only the most exceptional – best posture (literally and figuratively, i.e., those presenting a proper exterior aligned with the "reason" and values of the institution), fit bodies and minds – from previously excluded groups can be admitted. According to Alexander Weheliye, "hegemonic powers ... as a general rule, only grant a certain number of exceptions access to the spheres of full humanity, sentience, citizenship, and so on. This in turn, feeds into a discourse of putative scarcity in which already subjugated groups compete for limited resources, leading to a strengthening of the very mechanism that deem certain groups more disposable or not-quite-human than others" (2014, 13–14). Here is where we notice the politics of indifference in its most heightened state. Just as in the past where Black and Brown bodies and the bodies of women were used as evidence against them as scientists reported the so-called facts indifferent to these bodies as living or formerly living beings and simply specimens, so too with diversity and inclusion initiatives, the body provides evidence of itself as diversity is measured.

Ahmed writes, "When you embody diversity, you fulfil a policy commitment. Your body becomes a performance indicator. You become a tick in a box; you tick boxes" (2018, 335). And there are "specific ways in which 'diversity' over-counts while it subordinates" (Alexander 2005, 140). Today the workings of racism and sexism in the "inclusive" university have become ever more complex, nebulous and pervasive. And through the implementation of diversity and inclusion which "operates in apolitical and often antipolitical ways to selectively usher a few bodies into exclusive institutions" (Nash 2019, 24) the university puts these marginalized and minoritized bodies to work, doing the inclusion work that ensures its futurity.

The implementation of inclusion and equity strategies on campuses nationwide by individual institutions therefore is part of a larger systematic capitalist functioning in our postmodern societies where the workings of racism, sexisms and other 'isms on university campuses are pervasive and insidious. Even well-intentioned programs at the heart of our equity and inclusion initiatives have the impact of oppressing those whom they are supposed to serve, "requiring that the student decode forms of academic practice that are granted legitimacy through disciplinary technologies of assessment, ranking and measurement" (Burke 2018, 370). According to Desai and Murphy, diversity in its current form at the university "is preoccupied with the biopolitical management through technologies of counting and tracking minority student enrollments" (2018, 33). In actuality, the focus of equality and recognition programs based on narrow inclusion frameworks both hinders opportunities for real racial progress and creates conditions for the further perpetuation of inequality. According to Weheliye, "[i]f demanding recognition and inclusion remains at the center of minority politics, it will lead only to a delimited notion of personhood as property politics ... allowing for the continued existence of hierarchical differences between full humans" (2014, 81).

Understanding how today's inclusion as property politics reinscribes historical notions of bodily ownership of women and Black and Brown people by white men is of even greater concern today as we are also enduring rampant and pervasive "virulent gut patriotism ... [a] populist mobiliser – in part, because it feeds off the disappointed hopes of the present and the deep and unrequited traces of the past, imperial splendour penetrated into the bone and marrow of the national culture" (Hall 2017, 205–206). The imperialist (utopian) nostalgia for "the good old days" (e.g. make America great again) is at the heart of neoconservative, very thinly veiled ethnic nationalist arguments against diversity, as diversity itself is blamed for the fall of an idealized perfection in the university and the nation. Politics of resentment are wrapped up in the university's economic dependency on maintaining the comfort zone and not becoming the target of complaints by white conservatives with economic and culture-war power, and the collective identity as "normal" and "fit." The newly included find themselves caught in the middle, grinning and bearing.

When we think about inclusion into the university in relation to scarcity, we can better understand Nance's compulsion to write her article. Thinking about inclusion through the lens of scarcity helps us to go beyond the obvious of Iris Marion Young's politics of difference, as it is clear that even when, as she suggests, group difference is recognized and revalued, the apathetic methodology of power operates to further disenfranchise and marginalize groups considered different while appearing not to. We must keep this distinction in mind, for as difference continues to be accommodated through diversity and inclusion programs which target the previously excluded through using the same definitions that caused their inclusion, pain, and suffering in the first place as the metric to grant them access (Weheliye 2014, 75), their difference becomes their identity, the thing which the university recognizes, reacts to, and engages to further exploit them as it seeks its own advancement. The bodies of students and faculty who exist in this accommodation, i.e., *in* difference will never be considered normal, as difference is made to stick to, and live on and in them. They are expected to be grateful and not challenge issues important to them because they have been accommodated. As the institution makes accommodations for Black, Brown, and women students and faculty, the assumption is that they have done their part for the good of diversity through these performances. Because of this stance, they often become indifferent to requests from these students and faculty to engage with issues of inequity. The results are the continued perpetuation of the politics of indifference.

Between 2015 and 2018, I had a chance, as one of the few black women faculty at a small private liberal arts college, to witness first-hand how Black and Brown students (already socialized to survive a racialized and gendered history that constructs them as society's other) interacted with community cultures that construct them as strong and resilient (i.e., having the ability to overcome every adversity), and navigated the institutional barriers they came face to face with at this PWI. I observed the concrete ways in which historical, cultural and institutional barriers all worked together to marginalize these students, the various methods they employed to try to navigate their marginalization once enrolled, the toll that it took on their daily lives, and how difficult it was for them at this elite college to take care of themselves or to seek out care. During that time, I interviewed a total of 66 people, 41 of them Black and Brown students. Most of the students I interviewed admitted that they had been less prepared for college, both academically as well as culturally, than their white peers, and that this understanding impacted how they navigated the campus. This admission is in line with the research of Black feminist such as Venus Evans-Winters (2019). All of them spoke about their financial need to afford to be in college. For example, one of the research participants for this study, Gabby,[3] a third-year Latina student who walked approximately two miles to and from campus to work at the local McDonalds, stated in our interview, "only white students have hammocks or can play Frisbee on the Green[4]. I wish I was so lucky that I could just take time out to relax and play like them. Most of us [Black and Brown students] aren't." I took Gabby's comments to mean that Black and Brown students did not have the luxury to enjoy a "normal" college experience. Her mentioning the Green is an indication of how the campus itself was spatialized such that her white peers could occupy the center of campus, an observation harking back to McKittrick's point made earlier that space is socially constructed. In recognizing that the center of campus was not a place she could occupy, Gabby, as she watched white students make the campus their playground on her way to and from work, wished she was so lucky, she wished that she could relax and play. It is obvious from Gabby's words that watching these white students play provoked a set of specific feelings, she felt the campus differently from the playful white students.

Most of the students I interviewed for this study came from low-income families and are so-called first-generation students. I say so-called here because, as Evans-Winters notes, these students carry with them "the cultural legacy and knowledge of generations. Some students may be the first in their family to physically step foot on a college campus, or they are the *first in their family to attend college*, but that does not mean it is their first time learning; or that their appetite for learning or studying different cultures suddenly started when they entered that college space. This language, like racialized language, erases students' culture and humanity" (2019, 48). These students reported finding it extremely difficult to navigate their college environment, and didn't know how or whom to ask for help. Some of them came from underfunded inner-city schools, and were recruited through the Chicago-based Posse program,[5] a program which, while it provides opportunities for Black and Brown students, is an organization to which elite colleges outsource their diversity recruitment work. This outsourcing is part and parcel of how these neoliberal educational institutions promote a commitment to diversity while still being very much invested in ensuring the preservation of the status quo through a capitalist outsourcing and vetting process. This process takes the selection process out of the university's hands, making it someone else's responsibility, such that when students chosen via this method become or create crisis, the university can rid themselves of the student/crisis easily because they didn't actually pick these students when it comes down to it.

Some of the students whom I interviewed for this research shared that they attended schools where many of their peers had been pushed out or had gotten caught in the school-to-prison pipeline (Morris 2016). As they became the ones their friends and family felt had made it, they felt the added responsibility to represent their communities and families favorably. All of the students I interviewed for this research worked full or part-time, sometimes taking multiple jobs. Even those with a full (mostly Posse) scholarship had to pay hidden fees for books and other incidentals which those scholarships did not cover, and sometimes worked in addition to send money back home to help with various and mostly unforeseen family expenses. Many of the students interviewed were also very politically as well as socially involved in multiple groups for the promotion of racial and gender justice on campus, and felt like it was their responsibility to do the work of advocating to the administration for resources that could help their communities while they were still in college.

During the time I worked at the institution where this study took place, the administration announced on several occasions that it was working to provide the structural accommodations and resources necessary for students of color to thrive. For example, they sought to implement a program of full participation, another vague and empty diversity term which focuses on creating institutions which enable people from various backgrounds and with various identities throughout the institutional hierarchy to engage in meaningful ways to institutional life and to "contribute to the flourishing of others" (Strum et al. 2011). In the drive to implement full participation, the institution put together a committee, and conducted interviews and workshops to produce a report on the structural and financial barriers that existed on the campus. However, as Sarah Ahmed writes, this work to gather what she refers to as "'perception data,' that is, data that is collected by organizations about how they are perceived by external communities" and used to improve and inform their diversity work, "becomes about generating the 'right image' and correcting the wrong one" (2012, 34).

According to Ahmed, diversity therefore "becomes about *changing perceptions of whiteness rather than changing the whiteness of organizations.* Changing perceptions of whiteness can be how an institution can reproduce whiteness, as that which exists but is no longer perceived" (2012, 34). As such, these initiatives can produce results that superficially signal success in doing the work of diversity – such as statistics of increased enrollment, recruitment and retention of students that fill their diversity quota – while also using the labor (sometimes unpaid and unrewarded) of said students who have to fight for scant resources. As Chandra Mohanty writes, what the institution defines as diversity "bypasses power as well as history to suggest a harmonious empty pluralism" (2003, 193). These institutions engage in a diversity performance that is about measuring and marketing to a consumer market (Mohanty 2003, 141) rather than in a true commitment to equity. According to Salvador Vidal-Ortiz in their 2017 essay:

In the view of corporate-minded academic administrators, the more diversity there is, the more "experience" students gain. That, in coded language, means more diversity allows U.S.-born, non-Hispanic white students to consume otherness and develop the appropriate skills at

managing difference (and a portrayal of their "tolerance" for difference) for when they work with – not *in* – a "diverse" environment.
This translates in a direct gain – monetary and otherwise – for a white
student body that eventually becomes part of the work force. Their
coded experience with "diversity" allows them to "manage" diversity
without having to address inequality. It also means disciplining students of color to assimilate to that diversity project – preparing them
to abide by these unequal work-force standards, to fit within that system (2017, n.p.).

The impact of this performance is evident, as a majority of the students
I interviewed for this study shared many instances of microaggression
and outright racism on campus, and most felt that despite institutional
rhetoric and busy-ness, no real changes were being made. Many reported
feeling as though they did not belong, expressing feelings of frustration,
depression, helplessness, and that they just needed to get their degree and
get out.

In her 2016 text *In the Wake: On Blackness and Being*, Black feminist
Christina Sharpe describes what she calls "the weather," that is, pervasive anti-blackness, "*as* climate." Sharpe states "[t]he weather necessitates
changeability and improvisation; it is the atmospheric condition of time
and place; it produces new ecologies" (106). I use Sharpe's analogy of the
weather/climate here to describe how the weather/climate on our campuses –
campus climate – creates atmospheric conditions that can produce an ecology where Black and Brown students find themselves out of sync with their
physical surroundings and thus begin to enact a form of dissemblance.
Darlene Clark Hine (1989) describes the culture of dissemblance as the
way in which Black women conceal their thoughts and emotions to protect themselves from systems that demand their happy acquiescence while
working to oppress them, or what Black women call "going along to get
along." During the study, it was evident that the students engaged in a culture of dissemblance because they understood that the way they felt the
university was in most cases at odds with how they were to demonstrate
what they felt. The students, faculty and administrators I interviewed for
this study all suggested that Black and Brown students actively worked to
hide their authentic feelings by pretending to be okay even when they were
not. Hall and Sandler (1982) also write about what they call the "chilly climate" to capture ways in which institutions are open to addressing issues
of race/ethnicity, sexual orientation, gender, etc. This type of anti-black
weather/climate produces what Arline Geronimus refers to as "weathering." Geronimus, writing specifically about maternal mortality, states that
we need to be attentive to "the ways in which social inequality, racial discrimination, or race bias in exposures to psychological or environmental
hazards may ... affect differently the health of black ... women" (1996, 590).
Other scholars have built on this hypothesis and applied it to other aspects

of Black women's lives to show that racial inequalities have an impact on their health (Geronimus et al. 2006).

Being weathered or weathering produces, I argue, a student body "that submissively stays in place" (McKittrick 2006, 9) through suppression of not only the physical but also of the emotional. Being weathered is to be stripped; frayed like a tree or structure that stays in place and continually takes a beating from the weather. To be weathered is to be battered by the force of a system, enough to be dispossessed of the energy or the where-withal, while still being connected to an ecosystem that requires you to show up in ways that demand you perform, so as to keep up the inclusive perception. As Black and Brown students continue to be pounded by the climate produced by the neoliberal capitalist academy – in the form of get-ting good grades, smiling for the glossy catalogs, worrying about student loan obligations long before the debt would come due, the pressure to grad-uate on time and get a job – the anxiety of dealing with these things which on the surface would seem universal to all college students carries what Geronimus describes in terms of the "allostatic" load; this I argue, has a different type of weathering effect on them than on white students who typi-cally have shelter and are "sheltered" from the climate. White (middle class) students literally do not have to stay in place, as they leave campus for home and vacations during spring, summer, fall, Thanksgiving, and Christmas breaks, while some of the students I interviewed have to search for employ-ment on campus during breaks, not able to afford the cost of traveling home, and also to offset their living expenses.

Reading the university, as they must do in order to stay on top of the weather, foregrounds the ways in which Black and Brown students feel the university as they show up as numbers and bodies measuring the universi-ty's diversity. This reading practice brushes up against the official diversity numbers and the smiling photographed faces, to, as Hartman states, "read against the grain ... as a combination of foraging and disfiguration – raiding for fragments upon which other narratives can be spun and misshaping and deforming the testimony through selective quotation and the amolification of issues germane to this study" (1997, 12). That is to say, they read to make visible not only the smiling but the "grinning and bearing." To describe this process, I will now focus on the narrative of one particular student, a Black woman, to provide a space for her affective experiences at the university. I present her story on its own, for, as M. NourbeSe Philip states in an inter-view with Patricia Saunders, in response to a question about being "in front of this whole big mass of history, ... [the] impulse ... to linger and ponder about "one meager girl ... the pause is itself a way of honoring her life ...". Philip says:

> I think it [pondering about the life of this one girl] is totally subver-sive in the face of the kind of broad-brush brutalizing where people just get reduced to Negro man, Negro woman, and ditto, ditto, ditto.

You pay attention to one, and it is such an amazing act—and one that spills over to all the other dittos—paying attention and taking care with just the one. Because that's all we can do is care one by one by one. And that's why it was so important for me to name (Saunders 2008, 77–78).

Sharpe too writes about the ways in which Black scholars "get wedged in the partial truths of the archives while trying to make sense of their silences, absences, and modes of dis/appearance. The methods most readily available to us sometimes, oftentimes, force us into positions that run counter to what we know ... [such that] we are expected to discard, discount, disregard, jettison, abandon, and measure ways of knowing and to enact epistemic violence that we know to be violence against others and ourselves" (2016, 12–13). As "the archive lies as it tells a truth" (McKittrick 2014, 22), it is a form of empirical dismemberment.

Using the same Black feminist praxis of care espoused by Hartman, Philip, and Sharpe, I engage with a reading of the archives inherent to Black feminist scholarship which weighs evidence drawn from empirical and experiential knowledge similarly (Harris-Perry 2011, 173). To do so, I engage deeply with the single narrative of this Black woman interview participant, pausing purposely to listen, not from across the research/researcher distance but to hear her, as Perry mentioned in Chapter 2 writes, passionate utterances as an intimate gesture toward recognizing her humanity. I take the time to be attentive to the ongoing violence "within the context of the contemporary university where 'diversity' is tokenistically but not substantively prioritized, [such that] racialized and gendered management ... occur[s] ... through a form of valorization and fetishization, albeit of a limited and facile type" (Hong 2008, 101).

In writing about the Haitian earthquake of 2010, Sharpe writes how she was attentive to the violences that

> deposited that little girl there, injured, in this archive, and the violence in the name of care of the placement of that taped word on her forehead, and then I kept looking because that could not be all there was to see or say. *I had to take care.* ...I was looking for more than the violence of the slave ship, the migrant and refugee ship, the container ship, and the medical ship. I saw that leaf in her hair, and with it I performed my own annotation that might open this image out into a life, however precarious, that was always there. *That leaf is stuck in her still neat braids. And I think, Somebody braided her hair before that earthquake hit* (2016, 120).

In addition, according to hooks: "When our lived experience of theorizing is fundamentally linked to processes of self-recovery, of collective liberation, no gap exists between theory and practice. Indeed, what such experience makes more evident is the bond between the two – the ultimately reciprocal process wherein one enables the other" (1994, 61). In the spirit of these liberatory scholarships, I pay this kind of purposeful attention to the ways in which the everyday impacts our lives, detailing this one interview among the 66 that really got me thinking about the power of the politics of indifference, pausing long enough to honor the feelings and emotions of this one Black student in order to honor the lives of others.

I interviewed Clare,[6] a young Black woman of Nigerian descent who at the time was in her sophomore year. Clare identified as cis female, Christian and working-class. Both her parents were born and raised in Nigeria, her dad from Lagos and her mom from Ibadan. Clare, whose native language is Yoruba, was born in Nigeria and moved to the United States with her dad and her brother when she was five years old. Her mom followed later. Her parents finally settled in Chicago and eventually Clare was able to attend this high price tag, elite college through a Posse Scholarship which covered her full tuition. She said she chose this particular school because a student from her hometown who went to her high school was attending the college and told her he thought she would thrive there. She said she also wanted to attend because she understood that in the real world, she was not "always going to be put in situations with people of my same skin tone, same values, same culture, all that stuff. So I might as well learn now while I can instead of being thrown into it later and not know what to do." Clare's words indicate that not only is she preparing herself for a better future and everything one wants in a normal college experience, but that she is also practicing for a future when the stakes of being different might be higher.

When I spoke with Clare, she seemed to be thinking a great deal about how she fits into the academic space which she inhabited. She shared how student life at the college to date was both an academic and social struggle.

> The first semester here I went through a time where I felt like I didn't belong here or couldn't make it here because umm, I was taking a lot of science courses where I was the only woman of color or the only African-American woman in the class, and that was just wow so do I belong here? Obviously I stand out, but at the same time, standing out, I was put in the corner. I was a shadow .... You know, being around students who went to high schools with more funding, more money, all those things came at once because I thought I was prepared for the culture shock coming here and I was not. Umm, on the social side, same thing. I don't think there are a lot of things on campus that umm allows me to express myself the way I want to, umm I am not the type of person that wants to go out every weekend, go party every weekend, so that

definitely, and that is the big culture here so that definitely doesn't help me personally. It's been tough, it's been rough. Umm feeling like 1) I don't belong and 2) Not many people understanding me. And the ones that do understand me are upperclassmen who are about to graduate and leave so also having to get used to that.

Later in the interview, Clare reiterated this point when she said, "I feel like when I am here, I am a visitor in someone else's home." According to Puwar, these types of institutional spaces designate "through processes of historical sedimentation, certain types of bodies ... as being the "natural" occupants ... hav[ing] the right to belong ... while others are marked out as trespassers who are in accordance with how both spaces and bodies are imagined, politically, historically and conceptually circumscribed as being "out of place"" (2004, 51). Puwar's statement about historical sedimentation resonates with those references earlier by Stuart Hall about the "imperial splendour penetrated into the bone and marrow of the national culture." And this, no doubt, can be felt by those who are marked as out of place. Clare enumerates a number of reasons why she feels like a visitor including microaggressions (Sue et al. 2007), visible racial divisions on campus, and marginalization.

> You know those moments like am I crazy, am I making this up? Like that sorta thing. Uhh she [white roommate's white friend] walked in and she said, "Oh wow it smells like burnt hair and chicken in here," and hahaha and I was just first of all, my roommate was very uncomfortable, which I guess it is good that she didn't feel comfortable in that situation and try to like laugh it off, and I was there like umm did I just hear that? Like wow. So I had to go tell my friends and like yea you are not crazy, this is crazy. They couldn't believe someone said that. And I just shut down. I shut down when I see her on campus. I shut down for a while after that semester and I was like yo, someone actually said it smells like burnt hair and chicken in here. First of all, my hair was in braids. There wasn't any, any I don't know. Yea.

Microaggression scholars such as Sue and others have coined the term *attributional ambiguity* (Sue 2010) to describe the uncertain ways of navigating microaggressions including the questions people ask themselves as they try to unpack whether they have experienced a microaggression. Questions similar to Clare's, "Did what I think happened, really happen? Was this a deliberate act or an unintentional slight? How should I respond? Sit and stew on it or confront the person? If I bring the topic up, how do I prove it? Is it really worth the effort? Should I just drop the matter" (Sue et al. 2007, 279). So that not only is the person in a position of having no solid map to the territory of microaggression, having to question whether they are a reliable assessor of their own experience, they are also having to determine whether even mentioning it would be perceived as valid.

With regard to invisible racial divisions, Clare says in response to my question about navigating space as a Black woman on campus,

> The simplest example I can give is at the cafeteria. If you walk into [the Cafeteria], you can definitely tell that there is a divide. There is one side that is untouched by people of color and there is obviously this other side that all the people of color hang out and use. There is definitely a divide in clubs that pertain to talking about race and identity. Most of the attendance is people of color umm umm you can … I saw more of a divide after the election [2016], which was really sad. Those are the times when you think people come together umm yea there is definitely a yea. Yes.

This experience which is very familiar to Black and Brown students was explicitly named by sociologist Beverly Daniel Tatum in her seminal text, *Why Are All The Black Kids Sitting Together In the Cafeteria?: And Other Conversations About Race*, in which she writes about contemporary forms of racialized and class segregation (2017). Experiencing this segregation at institutions of higher education can be disorienting. As Heidi Safia Mirza writes, "there is 'disorientation', a double-take as you enter a room, as if you are not supposed to be there. You are noticed and it is uncomfortable" (2017, 44). This disorientation is certainly ironic as the introduction to college life is called "orientation." For students entering a space where their race marks them as a body "out of place," homing, place-making and place-finding are fraught with ambiguity and having to navigate and in effect "orient" to a de-centering "norm" which locates them outside "the normal college experience."

Regarding marginalization, Clare prefaces her comments saying she recognizes that non-white people are not a monolithic group, saying:

> some of the people of color that end up on this campus are, their background in dealing with situations like this is better than others. Umm a lot of people of color that I've talked to who like you know get into this environment and keep their cool is because their high school or previous educations was very similar to this. It looked like this. They were probably one of the fewer people of color on their campus. So they have figured out a way I guess to work that system and so that that umm usually when there is conflict dealing with color, they are never there because for some reason, I don't want to say white-washed, but…There are a couple of people, a couple of people who are white-washed. And I think a lot of it has to do with like their background where they were raised. If we broke up the stats, because they have stats of people of color on this campus. If you broke up the

stats to see people of color who grew up in urban cities and went to public schools, that percentage would be very small. And I don't know if that is on purpose or not. That is why I say maybe. Or maybe it just happens to work that way.

And so here we find Clare feeling the university, and by feeling it I mean, creating her own experiential metadata out of what the university would in fact describe very simply as a monolithic measurement of Black numbers. Speaking specifically about the marginality of Black and Brown students Clare observes:

When it comes to having events on campus, a lot of things about those events are not inclusive. For instance ... [at a] festival [on campus], a lot of the music was geared towards a white audience. Like I remember most of the people of color that day were just standing on the side when most of the students were having fun. Talking to the students here, programs, like Africana Studies, the amount of funding that gets, versus a philosophy or science. Which is here predominantly dominated by white people versus Africana which is predominantly dominated by people of color. Umm, yea. Stuff like that definitely you can see the divide.

Several scholars have written about the marginalization of courses relevant to students' historical and cultural backgrounds, experiences, and interests, as well as the lack of resources made available specifically for them, including student housing, centers, networking opportunities, and mental health resources (Hundle 2019, 314-315). This marginalizing and lack of resources is, as mentioned earlier, an orientation to campus life that is disorientation away from an environment which is made more for them and their unique needs, social, academic, and pre-professional.

Clare also shared how she felt like she was constantly under surveillance: "not that anyone in specific is watching, like the other students, or professors, or even the dean, but that like umm, everyone is watching." In listening to Clare describe her sense of being watched, I am reminded of what Patricia Hill Collins refers to as "the politics of containment," wherein Black women are under constant surveillance in the academy (Collins 1998). This surveillance is meant to circumvent any presumed illegitimate behaviors as simple as meeting together to engaging in demands for justice in the form of peaceful protest, and to restrict their movements, to depoliticize their oppression and to dissuade their resistance so as to keep them in their place at the bottom. Black women are aware that they are constantly being watched and placed under what Brittney Cooper calls a "hermeneutic of suspicion" (2018, 262) such that Black women often police themselves and their behavior. Students feel watched and constrict themselves so as not to be seen as a problem as a crisis in the way that

Crawley mentioned previously means. They engage in their own forms of "crisis management."

In fact, according to Penny Jane Burke, "inclusion often perpetuates problematic deficit perspectives that place the responsibility on those individuals who are identified as at risk of exclusion through their 'lack' of aspiration, confidence, adaptability or resilience. Inclusion might also be seen as a discursive space in which the politics of shame play out in ways that are experienced as personal failure and simply not being the 'right' kind of person and worthy enough for participation in higher education" (2018, 371). Inclusion becomes their burden to bear, and a heavy one at that, where students have to perform that diversity role expected of them (Ahmed 2018, 338), in that their presence provides the university with its much-coveted but unwanted diversity – and to provide even that presence requires labor (Ahmed 2018, 333). These students are on the clock, monitored, managed, and even guilted and shamed into "stepping it up lively." Clare, for example, lists the amount of work she does on campus not connected to her studies.

> I am a part of the club UMOJA on campus. I am on the e-board so, planning stuff for this month. February is definitely part of my job. I am also a part of SOAR, Students Organized Against Racism on campus. CSA, umm yea, uhh Reflections, which is the spoken word group on campus. Yea so those outlets. ... I go to the local ... library to tutor some students. ... back home I was a part of a tutoring program in which uhh in which umm you were paired up with a tutor who was essentially your mentor. I was umm in the program you were paired up with a tutor who was essentially a mentor to you and I think that program. I was in that program for about 5 years. I think that program definitely shaped how I think about mentorship and having someone in your corner. And umm I'd never had the chance to give back to that specific program because I am out here but I thought maybe doing something like that out here would suffice.

Most of what Clare describes above was also recounted to me by other students in most of the interviews I did. Students told me about a type of *thickness* they have to wade through – this dense weather where they were "getting, losing, and keeping their bearing within a thick present" (Berlant 2011, 198). This weather, according to Christina Sharpe, "is not the specifics of any one event or set of events that are endlessly repeatable and repeated, but the totality of the environments in which we struggle; the machines in which we live ... the weather" (2016, 111). This weather necessitates adaptability, "changeability and improvisation; it is the atmospheric condition of time and place; it produces new ecologies. *Ecology: the branch of biology that deals with the relations of organisms to one another and to their physical surroundings; the political movement that seeks to protect the environment, especially from pollution."* (Sharpe 2016, 106). In order for Clare to navigate the campus climate, she has to adapt both academically and socially.

It is ironic how academic institutions conduct and publish campus climate surveys to measure and monitor the perception of, as Sharpe calls it, "plantation management" where "[a]wareness of the ecological systems was necessary for the growth and cultivation of certain groups" (2016, 112). The intentionality behind these surveys is never about the minoritized and marginalized but rather about managing them and any potential crisis that may arise because of their presence on these campuses. Ashton Crawley asks us, what do we do "when our presence is crisis, when we carry crisis in the flesh" (2018, 12)? To think about the weather and ecology as Sharpe theorizes helps us to think about how these institutions are indifferent to how students like Clare experience "living in the wake of such pseudoscience, living the time when our labor is no longer necessary but our flesh, our bodies, are still the stuff out of which "democracy" is produced" (2016, 112). Clare's flesh as present and existing on a hostile campus is necessary to the survival of the university, operating in the wake of its racist scientific past, in which at the same time it is required to engage in inclusive gestures.

Throughout our interview, Clare kept reiterating the felt difference. The way in which she felt the university as a Black, working-class student at a predominantly white institution where there was a noticeable class difference between her and her white peers. In the below exchange, it is evident that Clare struggles with how she is made to show up as the university's difference and yet navigate that climate almost entirely on her own embodied as that difference.

Clare:	I am from Chicago and Chicago is kinda diverse. So this is not my first encounter with white people, but this is my first time seeing this attitude or type of culture or sense of community in a sense. That part I am not used to. That part was surprising.
ANB:	And by attitude and culture, you are referring to?
Clare:	This lack of community.
ANB:	Lack of community?
Clare:	On this campus. Yea.
ANB:	Umm which manifests in the [cafeteria] situation?
Clare:	Yes.
ANB:	How does this make you feel about being on campus or, you know, just being here?
Clare:	Umm it makes me sometimes feel like an animal. I won't lie. I am of a darker skin, which I definitely have gotten looks or stares not so friendly because of that. Especially when I have my natural hair out. This is a completely different person or thing to people and that, that is very scary sometimes. Umm yea.

That Clare uses the word "animal" to describe how she feels harkens back to the historical legacy of the university in justifying the nonhuman nature of Black people and non-personhood of women which still lives on in how this Black woman was feeling the university in 2018. We continue:

ANB: How do you feel about being a student of color on a predomi-
          nately white campus?

Clare: Umm I hate it. It's one of the hardest that I think anybody in col-
          lege has to do. Not obviously do you have to graduate, but then
          you have to figure out a way to not lose your head and lose your
          cool sometimes. You have to understand that not everybody
          understands you and who you are and has enough knowledge
          on what you face because of your skin tone or your skin color.
          And then, you can't get mad at them. You can't really get mad
          at people who don't know or do stuff or hurts and affects you
          personally. So you have to deal with more things than they do on
          campus. And I feel like when I am here, I am a visitor in someone
          else's home. Visitor in someone else's home and sometimes I feel
          I am here so the school can say they are inclusive. So it's like an
          institution is auctioning off my face to say that, "Look. Look at
          the black girl smiling. She loves it here. They all love it here."
          And that is very... For once that camera is off, the picture on the
          school website with the black girl with the Asian boy with the
          White girl walking. Once seen and over with, in reality, no one
          knows who you are. No one knows how you struggle here and
          how you have to keep your cool and keep your head above water
          every single day. Because you don't know what you are going to
          expect. You never know if today everyone is your ally. If tomor-
          row somebody wants to hurt you. And so they have to survive.
          To do that.

Clare's words perfectly "capture the affective, emotional, subjective and lived experiences of misrecognition and misrepresentation, that are felt in and through the body as forms of symbolic violence and injury on the self" (Burke 2018, 369). What we learn from Clare's single story reveals, as Hartman writes, "more about the horrors of the institution than did volumes of philosophy …the politics … and performance when dissolution and redress collude with one another and terror is yoked to enjoyment" (1997, 35). Clare's comments bring this chapter back around full circle, connecting her comments on photography as part of the technologies of oppression to the experience of those forced to take posture portraits, where the body is used as proof, measured, cataloged, and managed in order to be included. Hartman's theorizing mentioned earlier allows us to link the contemporary indifference of universities to "a repressive problematic of consensual and voluntarist agency that reinforces and romanticizes social hierarchy" where, as in the master-slave relationship Hartman writes about, the marginalized and minoritized student and the academy "are seen as, if not peacefully coexisting, at the very least enjoying a relationship of paternalistic dependency and reciprocity" (1997, 52). While the whip is no longer the technology used to oppress, the indifference behind its use continues on and is manifested in our current iteration of the afterlives of slavery and in diversity and inclusion where the singing, dancing, enslaved body has been replaced by the students as the "happy symbol of diversity." Meanwhile, as Clare relates, Black and Brown students can't even dance because the music the student activities programs prefer to sponsor is for the majority, the white students whose parents bring the lion's share of the tuition money, never the music Black and Brown students collectively enjoy and celebrate to! Not to mention that in a predominantly white college town, the surrounding opportunities for entertainment are much the same.

The student is represented as having agency in so far as "what is 'consented' to is a state of subjugation …[and] the notion of the autonomous self-endowed with free will is inadequate and, more important, inappropriate … by emphasizing complementarity, reciprocity, and shared values, this hegemonic or consensual model … neutralizes the dilemma of the object status and pained subject … and obscures … violence" (Hartman 1997, 53). This, McKittrick writes, is an essential part of "a plantation logic characteristic of (but not identical to) slavery [which] emerges in the present both ideologically and materially" (2013, 3). The abovementioned situations do not even begin to scratch the surface of all the ways in which marginalized and minoritized students endure the university climate. Just existing in this carefully constructed space can be traumatic especially when, as Ahmed writes, "your experiences of the organisation are not happy. The smile you provide masks more than organisational failure; it can also mask your own experience of that failure" (2018, 333) and thus failure of diversity practices in the current climate of the corporatized university which, according to M. Jacqui Alexander, requires "abrogate[d] speech … [and the] psychic, material, and spiritual costs of being put on display" (2005, 120). And the diversity practices (and those they are supposed to benefit) are figured as "failing" because of the "diverse" students, faculty, staff – when in fact the diversity practices are very deliberately constructed by administrations and committees.

This type of violence stays with you long after you leave the university, long after you have endured achieving that for which you yearned. In 2019, the New York Times published an Op-Ed by Harvard professor Anthony Abraham Jack who, in recounting his time as a Black low-income student writes, "students of color and those from lower-income backgrounds often bear the brunt of the tension that exists between proclamation and practice of this social experiment. Schools cannot simply showcase smiling Black and Brown faces in their glossy brochures and students wearing shirts blaring 'First Gen and Proud' in curated videos and then abdicate responsibility for the problems from home that a more diverse class may bring with them to campus" (2019, n.p.). The university's politics, interests, and how it positions minoritized and marginalized students in relation to its power are what condition the distorted representation of these students as happy and smiling. Such representation, according to Hartman, "must be taken into account, even as one tries to use this testimony for contrary purposes" (1997, 53). Until that account is made, these students are made to survive "their baccalaureate years navigating campus hate speech, dodging constantly changing DACA rulings, and living with the real consequences of repealing affirmative action and Title IX enforcement" (Harris-Perry 2018, n.p.). To do so, they have to engage in what Ahmed refers to as "*institutional passing… the work you do to pass *through* by passing *out* of an expectation: you try not to be the angry woman of color, the trouble maker, that difficult person. You have to demonstrate that you are willing to ease the burden of your own difference" (2018, 338).

For Clare, it came down to two questions: "How bad do you want this education? Are you willing to sacrifice things that make you comfortable? Those are the two questions because ultimately those are the two things you will have to deal with on this campus. On a predominately white campus. And if they can answer those questions I think they will be fine," she said. According to Clare and many of the students interviewed, there is never the possibility of refusal to consent. In fact, "[r]efusals in the context of domination are seldom received kindly" (Alexander 2005, 130). I understand Clare's pragmatism just as I understand why enslaved people danced and sang, and why bourgeois women stood passively to have their nude and semi-nude photos taken; "[f]or what defines this pressing situation is the problem of living in the ongoing now of it. The enduring present that is at once overpresent and enigmatic requires finding one's footing in new manners of being in it. The haunting question is how much of one's creativity and hypervigilant energy the situation will absorb before it destroys its subjects or finds a way to appear as merely a steady hum of livable crisis ordinary" (Berlant 2011, 196). Therefore, my questions engage not only with the pragmatic but also with escaping the longue durée of this present situation, such that I ask, how do we break through this weather? How do we, as Nourbese Philip asks, find freedom within these limitations (Saunders 2008, 65)? If, as Weheliye writes, "demanding recognition and inclusion remains at the center of minority politics, it will lead only to a delimited notion of personhood as property politics … for the continued existence of hierarchal differences between full humans…" (2014, 81), then what should we be demanding? I agree with Crawley that we should be calling for the destruction of "the very choreographies of marginality and violence that emerge from geographies of normativity." This requires that we struggle against the hierarchies of oppression set in motion by settler colonialism and slavery "as not just a problem that arrives on campus but one in which campuses participated" (Crawley 2018, 10). The only way of resisting these mechanisms is to read them, name them, classify them in return just as the oppressed have been classified.

In the following chapter, I propose another *way/where*. I continue reading in the vein of Hartman, by attempting to "brush history against the grain," and building on Davies' and Rinaldo Walcott's concepts of *elsewhere* to *whatever* as a means of "turning to forms of knowledge and practice not generally considered legitimate objects ... or appropriate or adequate sources for history making and attending to the cultivated silence, exclusion, relations of violence" (Hartman 1997, 11). I turn to something and somewhere else because the present situation of this longue durée is not working for minoritized and marginalized students, who are made to feel the university in unhealthy ways. Like da Silva, in this present moment, I can hear loudly "Prospero's laugh as Caliban now rehearses his productive power by selectively reading his books" (2007, 184). And yet, there is hope as seen in the very poignant retort/read Caliban provides to/of Prospero who claims Caliban was incapable of speaking any recognizable languages before Prospero civilized him. Caliban says, "You taught me language; and my profit on't/Is, I know how to curse./The red plague rid you/For learning me your language" (1611)! What a read!

In the following chapter, I turn to histories, places, and materials which allow me to "illuminate the practice of everyday life – specifically, tactics of resistance, modes of self-fashioning and figurations of freedom – and to investigate the construction of the subject and social relations contained within" them (Hartman 1997, 11). In what follows, I engage in a reading practice which in current vernacular is broadened as "reading for filth" – seeing the ugly, seeing the dirt, and thus creating a space to act out that tactical language and practice. I do so toward what Sharpe describes as *aspiration*.

Notes

1. Perhaps ironically this moral panic followed the abandonment of severe corsetry, which did in fact deform internal organs.
2. More on this study can be found at Andrea N. Baldwin, Heidi Henderson, and Sangyoon Lee, "(Re)creating the Posture Portraits: Artistic and Technological (Re)productions of the Gendered (Re)presentations of Bodies at Institutions of Higher Education." *Body Studies Journal* Vol. 2, no. 8 (2020): 79–100.
3. All names have been changed.
4. A big open outdoor grassy space in the center of campus.
5. The Posse, was started in 1989 and "is rooted in the belief that a small, diverse group of talented students—a Posse—carefully selected and trained, can serve as a catalyst for individual and community development" (Posse). The program's goals are threefold; to "expand the pool from which top colleges

and universities can recruit … from diverse backgrounds"; to "help these institutions build more interactive campus environments so that they can be more welcoming for people from all backgrounds," and to "ensure that Posse Scholars persist in their academic studies and graduate" to become leaders in society (Posse, retrieved 2019).

6. Pseudonym.

References

Ahmed, Sara. *On Being Included: Racism and Diversity in Institutional Life.* Durham: Duke University Press, 2012.

Ahmed, Sara. "Rocking the Boat: Women of Color Ad Diversity Workers." In *Dismantling Race in Higher Education: Racism, Whiteness and Decolonizing the Academy*, edited by Jason Arday and Heidi Safia Mirza, 331–348. London: Palgrave Macmillan, 2018.

Alexander, M. Jacqui. *Pedagogies of Crossing: Meditation on Feminism, Sexual Politics, Memory, and the Sacred.* Durham: Duke University Press, 2005.

Barnard College, *The Mortarboard.* New York: 1972.

Berlant, Lauren. *Cruel Optimism.* Durham: Duke University Press, 2011.

Burke, Penny Jane. "Trans/Forming Pedagogical Spaces: Race, Belonging and Recognition in Higher Education." In *Dismantling Race in Higher Education: Racism, Whiteness and Decolonizing the Academy*, edited by Jason Arday and Heidi Safia Mirza, 365–382. London: Palgrave Macmillan, 2018.

Chatterjee, Piya and Sunaina Maira. "Introduction: The Imperial University." In *The Imperial University: Academic Repression and Scholarly Dissent*, edited by Piya Chatterjee and Sunaina Maira, 1–50. Minneapolis: University of Minnesota Press, 2014.

Chen, Yush-mei. "A Study of Physical and Medical Examinations Given at Wellesley College Hygiene and Physical Education," *Wellesley College Archives and Special Collections* (1933) Box 305, Wellesley College.

Clark Hine, Darlene. "Rape and the Inner Lives of Black Women in the Middle East." *Signs* Vol. 14, no. 4 (1989): 912–920.

Collins, Patricia Hill. *Fighting Words: Black Women and the Search for Justice.* Minneapolis: University of Minnesota Press, 1998.

Connecticut College, *Connecticut College Alumnae News* Vol. 6, no. 1 (1928). Available at: https://digitalcommons.conncoll.edu/alumnews/17/ (accessed 10 October 2018).

Connecticut College, *Connecticut College News* Vol. 16, no. 1 (1930): 10–14. Available at: https://digitalcommons.conncoll.edu/cgi/viewcontent.cgi?referer= &httpsredir=1&article=1000&context=ccnews_1930_1931 (accessed 10 October 2018).

Crawley, Ashton. "Introduction to the Academy and What Can Be Done?" *Journal of Critical Ethnic Studies Association* Vol. 4, no. 1 (2018): 4–19.

da Silva, Denise Ferreira. *Toward A Global Idea of Race.* Minneapolis: University of Minnesota Press, 2007.

Desai, Jigna and Kevin P. Murphy. "Subjunctively Inhabiting the University." *Journal of Critical Ethnic Studies Association* Vol. 4, no. 1 (2018): 23–43.

Elliott-Cooper, Adam. ""Free, Decolonized Education" – A Lesson from the South African Student Struggle." In *Dismantling Race in Higher Education: Racism,*

Whiteness and Decolonizing the Academy, edited by Jason Arday and Heidi Safia Mirza, 289–296. London: Palgrave Macmillan, 2018.

Evans-Winters, Venus. *Black Feminism in Qualitative Inquiry: A Mosaic for Writing Our Daughter's Body*. New York: Routledge, 2019.

Geronimus, Arline T. "Black/white Differences in the Relationship of Maternal Age to Birthweight: A Population-Based Test of the Weathering Hypothesis." *Social Science & Medicine* Vol. 42, no.4 (1996): 589–597.

Geronimus, Arline, et al. "'Weathering' and Age Patterns of Allostatic Load Scores Among Blacks and Whites in the United States." *American Journal of Public Health* Vol. 96, no. 5 (2006): 826–833, doi: 10.2105/AJPH.2004.060749.

Glick, Megan H. *Infrahumanisms: Science, Culture, and the Making of Modern Non/Personhood*. Durham: Duke University Press, 2018.

Hall, Roberta and Sandler Bernice. "The Classroom Climate: A Chilly One for Women?" *Project on the Status and Education of Women*. Washington, D.C.: Association of American Colleges, 1982.

Hall, Stuart. "The Empire Strikes Back [written in 1982]." In *Political Writings: The Great Moving Right Show and Other Essays*, edited by Sally Davison, David Featherstone, Michael Rustin and Bill Schwarz, 200–206. Durham: Duke University Press, 2017.

Harris-Perry, Melissa. *Sister Citizen: Shame, Stereotypes, and Black Women in America*. New Haven: Yale University Press, 2011.

Harris-Perry, Melissa. "What It's Like to Be Black on Campus Now." *The Nation* (2018). https://www.thenation.com/article/archive/what-its-like-to-be-black-on-campus-now/.

Hartman, Saidiya V. *Scenes of Subjection: Terror, Slavery, and Self-Making in Nineteenth-Century America*. New York: Oxford University Press, 1997.

Hong, Grace Kyungwon. "'The Future of Our Worlds': Black Feminism and the Politics of Knowledge in the University under Globalization." *Meridians* Vol. 8, no. 2 (2008): 95–115.

hooks, bell. *Teaching to Transgress: Education as the Price of Freedom*. New York: Routledge, 1994.

Hundle, Anneeth Kaur. "Decolonizing Diversity: The Transnational Politics of Minority Racial Difference." *Public Culture* Vol. 31, no. 2 (2019): 289–322. doi: 10.1215/08992363-7286837.

Jack, Anthony Abraham. "I Was a Low-Income College Student. Classes Weren't the Hard Part: Schools Must Learn That When You Come From Poverty, You Need More Than Financial Aid To Succeed." *The New York Times*, 2019. https://www.nytimes.com/interactive/2019/09/10/magazine/college-inequality.html.

Keating, Ana Louise. *The Gloria Anzaldua Reader*. Durham: Duke University Press, 2009.

McKittrick, Katherine. *Demonic Grounds: Black Women and the Cartographies of Struggle*. Minnesota: University of Minnesota Press, 2006.

McKittrick, Katherine. "Plantation Futures." *Small Axe* Vol. 17, no. 3 (2013): 1–15. https://read.dukeupress.edu/small-axe/article-abstract/17/3%20(42)/1/33296/Plantation-Futures.

McKittrick, Katherine. "Mathematics Black Life." *The Black Scholar* Vol. 44, no. 2 (2014): 16–28.

Mingus, Mia. "Moving Toward the Ugly: A Politic Beyond Desirability." Femmes of Color Symposium Keynote Speech, Oakland CA, 2011.

Mirza Heidi, Safia. "'One in a Million': A Journey of a Post-Colonial Woman of Colour in the White Academy." In *Inside the Ivory Tower: Narratives of Women of Colour Surviving and Thriving in British Academia*, edited by Deborah Gabriel and Shirley Anne Tate, 39–53. London: Trentham Books, 2017.

Moffett- Bateau, Courtney. "American University Consensus and the Imaginative Power of Fiction." *Journal of Critical Ethnic Studies Association* Vol. 4, no. 1 (2018): 84–106.

Mohanty, Chandra Talpade. *Feminism without Borders: Decolonizing Theory, Practicing Solidarity*. Durham: Duke University Press, 2003.

Morris, Monique W. *Pushout: The Criminalization of Black Girls in Schools*. New York: The New Press, 2016.

Nash, Jennifer C. *Black Feminism Reimagined: After Intersectionality*. North Carolina: Duke University Press, 2019.

Perry, Imani. *Vexy Thing: On Gender and Liberation*. Durham: Duke University Press, 2018.

Puar, Jasbir K. "'I Would Rather Be a Cyborg than a Goddess': Becoming-Intersectional in Assemblage Theory." *philoSOPHIA* Vol. 2, no. 1 (2012): 49–66.

Puwar, Nirmal. *Space Invaders: Race, Gender and Bodies Out of Place*. Oxford: Berg, 2004.

Saunders, Patricia. "Defending the Dead, Confronting the Archive: A Conversation with M. NourbeSe Philip," *Small Axe* Vol. 12, no. 2 (2008): 63–79.

Sharpe, Christina. *In the Wake: On Blackness and Being*. Durham: Duke University Press, 2016.

Slaughter, Sheila and Gary Rhoades. *Academic Capitalism and the New Economy: Markets, State, and Higher Education*. Baltimore: John Hopkins University Press, 2004.

Smith, Christi M. *Reparations & Reconciliation: The Rise and Fall of Integrated Higher Education*. University of North Carolina: Chapel Hill, 2016.

Strings, Sabrina. *Fearing the Black Body: The Racial Origins of Fat Phobia*. New York: New York University Press, 2019.

Sturm, Susan, et al. *Full Participation: Building the Architecture for Diversity and Public Engagement in Higher Education (White Paper)*. Columbia University Law School: Center for Institutional and Social Change, 2011.

Sue, Derald Wing, et al. "Racial Microaggressions in Everyday Life: Implications for Clinical Practice." *American Psychologist* Vol. 64, no. 2 (2007): 271–286.

Sue, Derald Wing. *Microaggressions in Everyday Life: Race, Gender, and Sexual Orientation*. New Jersey: Wiley, 2010.

Tatum, Beverly Daniel. *Why Are All The Black Kids Sitting Together In the Cafeteria?: An Other Conversations About Race*. New York: Basic Books, 2017.

The Day, "Connecticut College Now Distinctly Intersectional, Dr. Blunt Reports; Reviews Year and Tells of Its Needs." Dec. 4, 1931.

Vassar Historian. *Posture and Photographs*, Vassar Encyclopedia, 2005. Available at http://vcencyclopedia.vassar.edu/student-organizations/athletics/posture-and-photographs.html.

Vertinsky, Patricia. "Physique as Destiny: William H. Sheldon, Barbara Honeyman Heath and the Struggle for Hegemony in the Science of Somatotyping." *Canadian Bulletin of Medical History* Vol. 24, no. 2 (2007): 297–298. doi: 10.3138/cbmh.24.2.291.

Vidal-Ortiz, Salvador. "Latinxs in Academe." *Inside Higher Ed*, September 22, 2017.

Weheliye, Alexander G. *Habeas Viscus: Racializing Assemblages, Biopolitics, and Black Feminist Theories of the Human*. Duke University: Durham, 2014.

Wynter, Sylvia. "No Humans Involved: An Open Letter to My Colleagues. *Forum N.H.I Knowledge for the 21st Century: Knowledge on Trial* Vol. 1, no. 1 (1994): 42–73.

Young, Iris Marion. *Justice and the Politics of Difference*. Princeton: Princeton University Press, 1990.

4 *Reading* toward aspiration

A Black feminist shad(e)y theoretics and the politics of elsewhere and whatever

All antiblack misogynoirist cultural projections are due for a *read*.
　　　　　　　　　　　　　　　　　　　　　　　　– Lomax 2018, xvi

...conceive of feminism not primarily as a set of positions or doctrines but as a critical practice for understanding and working against gendered forms of domination and against the way gender becomes a tool of domination and exploitation. ...engage in this critical reading practice with the stories, events, and cases presented.
　　　　　　　　　　　　　　　　　　　　　　　　– Perry 2018, 6

My attempt to read against the grain is perhaps best understood as a combination of foraging and disfiguration – raiding for fragments upon which other narratives can be spun and misshaping and deforming the testimony through selective quotation and the amolification of issues germane to this study.
　　　　　　　　　　　　　　　　　　　　　　　　– Hartman 1997, 12

As a foundation of the theories and observations analyzed in Chapters 2 and 3, my extensive historical overview in Chapter 1 demonstrates that "the U.S. academy is an 'imperial university'" (Chatterjee and Maira 2014). As the working of imperialism by, for, and through the university has existed since its founding in the United States, despite the so-called inclusion practices of the twenty-first century, they have resulted in the exclusion of those historically evaluated as outsiders to the realm of the mind. Subsequently, I have explored how those now included or seeking to be included in the university *feel* it, as they are known and know themselves to represent the university's diversity, and how they are expected/required to act accordingly. My reading makes it clear that in our present iteration of the university, there continue to be "facades of inclusion without any change to dominant and unequal power structures or knowledge bases" (Sultana 2019, 35), and that those are facilitated through the use of technologies of oppression that are leveraged by the politics of indifference.

DOI: 10.4324/9781003019442-5

Many before me have theorized about the university's investment in, and commitment to these structures, and demanded that the university itself be abolished. Calls for abolition fall on a spectrum. For many abolitionists, those do not mean "abolition as the elimination of anything but abolition as the founding of a new society" (Harney and Moten 2013, 42). According to Crawley, "[t]o ask what can be done with and about the academy, the university, however, is *not* to ask if knowledge will end. Rather, we move toward the end of a knowledge production that produces *as it is produced by* modern Man, the genre-specific man and its overrepresentation" (2018, 11).

In this chapter, I build upon some of the ideas of university abolitionists to advance a Black feminist shad(e)y theoretics which combines a queer and diasporic reading practice with a view of moving toward what Christina Sharpe in her text *In the Wake: Of Blackness and Being* refers to as *aspiration*. Sharpe asks,

> "What is the word for keeping and putting breath in the body? ...What are the words and forms for the ways we must continue to think and imagine laterally, across a series of relations in the hold, in multiple Black everydays of the wake? The word that I arrived at for such imagining and for keeping and putting breath back in the Black body in hostile weather is *aspiration* (and aspiration is violent and lifesaving). ... *Aspiration* is the word that I arrived at for keeping and putting breath in the Black body" (2016, 113).

Aspiration also means hope or ambition of achieving something, and therefore calls forth the lifegiving worldmaking possibilities of yearning. Calling forth these possibilities depends, according to Walcott on the types of reading practices we engage, how attuned our reading practices are to the history of struggle such that they can mount a challenge to what we take for granted as knowledge, to elicit "other and different kinds of responses" (2003, 118). This reading practice gives the university the proverbial side eye.

The reading practice I engage in this chapter is both queer and diasporic. It is queer in the sense that Omise'eke Natasha Tinsley writes about queerness as "marking disruption to the violence of normative order and powerfully so: connecting in ways that commodified flesh was never supposed to loving your own kind when your kind was supposed to cease to exist, forging interpersonal connections that counteract imperial desires for Africans' living deaths" (2008, 199). Tinsley words resonates with Lorde's philosophical poetics, "we were never meant to survive" (1997). This queer reading practice is also similar to what Scott Bravmann refers to as "the critical lens of queer cultural studies of history that investigate the retelling of history as vehicles for mobilizing new social subjects, contesting hegemonic social definitions, and creating new cultural possibilities both in the present and for the future" (1997, 128). This queer reading works not only to interrogate but also to disrupt perceived academic normative structures (Miller and Rodriguez 2016, xvi).

The reading practice is also diasporic, as Walcott explains how we/the diasporic subject,

> informed by the peregrinations of ... an Atlantic consciousness ... can bear to tolerate that what is at stake in our reading of any text had much to do with our interactions within, across and outside our given localities, regions, nations and continents ... attempt to understand how national and outer-national practices and desires inform our readings. But equally important it is a reading which seeks to be transgressive in the contexts of all official readings of blackness. It is also a reading which is crucial to make sense of what (with short hand doubt) I call the diaspora queer speaker. It is above all a tentative reading (2003, 118).

It is akin to what M Jacqui Alexander means when she writes about a daily practice necessary to create shifts of perception so as to realize the possibilities of change. This practice writes Alexander, "is the *how*: it makes the change and grounds the work. A reversal of the inherited ... [it is] heart engaged action ... at the deepest, most spiritual level meaning in our lives. It is how we constitute our humanity" (2011, 80). A Black feminist shad(e)y theoretics is heart engaged, caring enough to account for the complex lived realities of the marginalized and minoritized beyond institutional readings, making it clear that official (historical/archival) accounts of Black and Brown people are in fact intentional misreadings.

In 1990 the documentary *Paris Is Burning* (Livingston 1991) was released. The film has gained notoriety and been talked about over decades by supporters and critics alike. It details the lived experiences of Black and Brown queer folx who took part in the drag ball scene in New York, co-creating what I am arguing was an *elsewhere* – in the way the Davies defines elsewhere. For those queer folx excluded from major societal institutions, this was literally and figuratively a place of aspiration. The film introduced us to several people prominent in the scene at that time (Dorian Corey and Pepper LaBeija, Venus Xtravaganza, Octavia St. Laurent, Paris Dupree, Willi Ninja, among others), and to several terms including what it means to *read*, to *throw shade*, and the concept of *realness*. As I have described in the Introduction of this book, in one of the most *legendary* scenes in the film (in the sense of its acclaim but also in the very terms of the drag ball culture and language – to be legendary is to not only be iconic but also a creator of *realness*, a living definer as well as definition), Corey explains these terms in detail in essence, documenting and making public what was previously an insider/outsider, orally transmitted form of cultural theorizing. Corey states, "Shade comes from reading. Reading came first. You get in a smart crack, and everyone laughs and kikis because you've found a flaw and exaggerated it, then you've got a good read going. Shade is I don't tell you you're ugly but I don't have to tell you because you know you're ugly ... and that's shade."

In this chapter, I describe how the reading I have been doing thus far in the text, while it has been able to demonstrate how the politics of indifference deployed through the apathetic methodology of power preys on the yearnings of those the university desires to make up their diversity quota, also as Corey states, advances to shade. The reading turns the gaze of the "outsider" excluded/included upon the institution, demonstrating that while the university has been able to continue an oppressive reign that they must be held to account and must also be made to "grin and bear it." This reading is a reversal of the inherited, it helps us to have, search for and "sense for otherwise possibility ... to make ourselves open and vulnerable to otherwise possibility while also protecting [us] from harm [by] the violence that attends being open and vulnerable" (Crawley 2018, 18). It permits us to "fail to live up to the university while showing up for each other ... [to engage in] intentioned failure to produce the politics of the neoliberal academy while concurrently practicing a poetics of relationality, sociality, fundamental connection, and disorientation and unsettlement" (Crawley 2018, 18). This reading that turns to shade becomes the basis for this Black feminist Shad(e)y theoretics which, as Corey states, doesn't have to tell the university it is ugly/imperial/oppressive because it already knows. But just like the subversive performances including the use of language depicted in *Paris Is Burning* by those oppressed and exclude from heteronormative society to make their own spaces elsewhere, this Black feminist Shad(e)y theoretics raises the proverbial middle finger to the university, turns away with a resounding "whatever!" and goes to the elsewhere, the space where Black and Brown people have always produced knowledge.

My Black feminist Shad(e)y theoretics also engages with queer theorist José Esteban Muñoz's work on disidentification. Muñoz acknowledges, much like Hartman's nuanced read on the dancing and singing of the enslaved, that "[n]ot all performances are liberatory or transformative. Performance, from the positionality of the minoritarian subject, is sometimes nothing short of forced labor ... Minoritarian subjects do not always dance because they are happy; sometimes they dance because their feet are being shot at" (1999, 189). Muñoz calls this "[t]he 'burden of liveness' ... a cultural imperative within the majoritarian public sphere that denies subalterns access to larger channels of representation, while calling the minoritarian subject to the stage, performing her or his alterity as a consumable local spectacle" (182). This reading of the burden of liveness evokes the burden of diversity which Black and Brown students and women are required to carry, as I have described throughout this text, particularly in Chapters 2 and 3; about the relationality between Black and Brown students, and women and the university and the perception of agency. Muñoz posits a theory of disidentification which he writes is a strategy "the minority subject practices in order to negotiate a phobic majoritarian public sphere that continuously elides or punishes the existence of subjects who do not conform to the phantasm of normative citizenship" (4). Muñoz's theorizing of disidentification as a strategy reminds me of Clare's final words of advice in our interview to students coming after her, which is: if they want it (this education) bad enough, they just have to continually negotiate a space that is harmful to them in order to survive. Muñoz describes disidentification as being "about cultural, material, and psychic survival. It is a response to state and global power apparatuses that employ systems of racial, sexual, and national subjugation ... managing and negotiating historical trauma and systemic violence" (161). He further posits that this strategy is "sometimes ... insufficient..." (162) – and I agree with this – however I still witness that this is exactly what students, in particular, Black and Brown students do to survive the university.

With Muñoz, I observe that disidentification is not only about survival but also a "powerful and seductive site of self-creation" (4) which offers the minoritarian subject the ability not only to "exist in the *moment* [but] the privilege or the pleasure of being a historical subject [and] ...the luxury of thinking about the future" (189). Muñoz's theorizing has been instructive to the ways in which I think about a Black feminist Shad(e)y theoretics as "[w]hen our lived experience of theorizing is fundamentally linked to processes of self-recovery, of collective liberation, [where] no gap exists between theory and practice" (hooks 1994, 61). It is about the *how* of politics such that it envisions the political value of shade as liberatory in contrast to the politics of shame, mentioned in the previous chapter, mobilized by universities to whip people into shape/line. Shade is a good read while shame is a misread. Having read the university, we understand that just as we know they are ugly, the institution knows it too, and yet they will not change because their ugliness is important to their continuation. Throwing shade at the university also means shading the body/mind duality invented by the academy to justify its racist, sexist and ableist underpinnings. It is to expose the affective and the ways in which the curation of the university as a place of knowledge and not of feelings is false. Yet, for me, the real potential of shading is not only to hold up a mirror to the university letting them know that we know they are ugly but also to allow us to explore places, spaces, times, things – *whatever else and elsewhere* – options beyond and other than the university as it currently stands.

Riffing off of Muñoz again, disidentification then contains the creative potential for "world making," which occurs when minoritarian subjects are "*not* content merely to survive, but instead to use the stuff of the 'real world' ... and willful enactments of the self for others ... as spectators and actors ... to continue disidentifying with this world until [they] achieve new ones" (1999, 200). As such searching for a *whatever* else and *elsewhere* of the university is not automatic but requires that we take the time to search and engage willfully and intentionally. Borrowing Muñoz's definition of world-making, I invoke the elsewhere and the whatever as spatio-temporal concepts that moves us toward futurity and that are very important to a Black feminist Shad(e)y theoretics because it demonstrates there is potential for worldmaking/creating an elsewhere even as we are preoccupied with dancing because our feet are being shot at.

I argue that Munoz's disidentification must be read alongside the concept of realness as explored in *Paris Is Burning*. In the film, many of those featured, particularly Venus Xtravaganza desired/yearned to perform realness as it would be recognized by white cis-heteronormative society. In the film, Venus Xtravaganza laments, "I would like to be a spoiled, rich white girl ... They get what they want, whenever they want it." While her particular yearning may not be the same as that of the aspiring scholar, Venus' aspiration offers a parallel regarding the politics of yearning and how it is important to the ways in which we not only create spaces of survival for ourselves in the present but also how we engage in visions of creating the future we yearn for somewhere other than the here and now of our oppression. And while these visions are often manifested through a yearning that is coopted and constrained by white, cis gender, heteronormative capitalist ideals, the political potential of this yearning is important to worldmaking, achieving an elsewhere and whatever. Yearning to achieve realness, some sense of normalcy, as we see in the film, results in all types of worldmaking potential in the ball scene – community, homes, mothering, voguing, etc., such that yearning and the politics of yearning is not inherently a bad thing. Even though yearning is usually tethered to a long history of capitalist formations and ambitions/aspirations to "make it" in the realms those formulations rule over, there are creative and worldmaking possibilities that comes from working to satisfy that yearning outside of institutionally preordained ways of university-sanctioned education. This yearning can be fulfilled through other means (whatever) and in other places (elsewhere). Imani Perry writes "There are many sorts of callings, yearnings, and hauntings. Some are terrifying; some are traumatizing. Some are mesmerizing. Avery Gordon describes a sort that makes us move toward *something* but also toward a different way of being. Among the hauntings lie the quest for meaning, intimacy, and jot, the fundamental desires that undergird so much of our curiosity and imagination" (2018, 199). Fulfilling yearnings beyond the institutions which the capitalist society has stipulated means that one can be as imaginative as one wishes, grabbing hold of whatever you can find, whatever time you have toward imagining *whatever* possibilities, and all types of *aspiration*. According to Kevin Quashie quoting

Black feminist Patricia Hill Collins' *Black Feminist Thought*, we should think of "imagination as the capacity to call one's world into being; it is imagining as an act of deliberateness and self-making. ...imagination is "consciousness as a sphere of freedom"" (2012, 43-44). Working to achieve the worldmaking possibilities starting from the elsewhere of the imagination, we can be whatever.

When we think of yearning in this way, as moving from our interior and as coming from a sense of who we are, we are confronted with the possibilities of what Audre Lorde terms the erotic. In *Uses of the Erotic the Erotic as Power*, Lorde writes "[th]e erotic is a measure between the beginnings of our sense of self and the chaos of our strongest feelings. It *is an internal sense of satisfaction* to which, once we have experienced it, we know we can aspire. For having experienced the fullness of this depth of feeling and recognizing its power, in honor and self-respect we can require no less of ourselves." [emphasis mine] (1984, 54). In Chapter 2 of this text, I introduced the promise of yearning and how institutions prey upon this promise. I suggest here that to engage with one's interior and imaginary, connects intrinsically to yearning where our desires are "not merely a naïve rejection of the realities of social inequality ... [but also] holds firm to the right to be human. In moving beyond the social and political implications of the body toward a celebration of spirit and feeling, ...not a consciousness that is doubled and encumbered, but a consciousness that is free, full of wander and wonder, where surrender – not resistance – is an ethic" (Quashie 2012, 32). When Quashie invokes surrender, it does not mean a surrender to university logics but rather a surrender to one's interior, and (following Lorde) the chaos of our strongest feelings. Imagine the types of disruption that would result if minoritized and marginalized students expressed how they really feel about being the university's diversity. This surrender would result in "disruption to social logics at the emotional register, a disruption that allows us to imagine differently" (Perry 2018, 214). Lorde writes that once we move from that place of the erotic whatever we do fulfils. This in turn leads me to Barbara Christian's words: "I can only speak for myself. But what I write and how I write is done in order to save my own life. And I mean that literally. ... a way of knowing that I am not hallucinating, that whatever I feel/know *is*. It is an affirmation that sensuality is intelligence, that sensual language is language that makes sense" (1987, 61). Christian's words resonate with the disorientation of students like Clare, who ask themselves, "Did this really happen?" and yes, it did happen, she did not hallucinate. The evidence of her eyes and ears and other ways of knowing tells the truth and she held to that toward the goal of survival and "getting out" – and Quashie, Lorde, and Christian offer up a beyond, not mere survival, in surrender to everything the knowing and feeling reveals the name of it, calls it out.

In the film *Paris Is Burning*, we are confronted with the stark and sometimes haunting lived realities of impoverished, Queer Black and Brown folx

and their longings, as well as the determination and precision of prepara-
tion and walking in the ball. We are also witness to many very intimate
and playful moments such as the scene on the pier when they are reading
each other and talking about their respective houses. One of the defining
aspects of the film for me is the ways in which the lives of queer folx featured are not flattened but that they are portrayed in their complexity. I
will be the first to admit that there is a lot about *Paris Is Burning* that is
concerning particularly that it was directed and produced by a cis gender
white woman so that the lens through which we are able to reach these
elsewhere communities is indeed mediated through this gaze. However,
the abovementioned *intimacy* and *playfulness* between the people this lens
shows us is palpable; and enough to lead us to the worldmaking potential
of an elsewhere. In her work "Playfulness, "World"- Travelling, and Loving
Perception," María Lugones specifically engages the concept of playfulness
as having the potential for worldmaking. She defines playfulness as "in
part, an openness to being a fool, which is a combination of not worrying
about competence, not being self-important, not taking norms as sacred
and finding ambiguity and double edges a source of wisdom and delight"
(1990, 400-401). Being playful then is the opposite of what the university
is. It is part of what the university defines as failure, unscholarly, unfit for
the serious "life of the mind" and knowledge production. To be playful
therefore can only be had as Lugones writes if we travel – if we go/look/
seek elsewhere, that is, "travel with the interstices of power and to peel
away the deposits that coagulated along the way" (Alexander 2005, 143).
She continues "Travelling to each other's "worlds" enables us to *be* through
loving each other" (1990, 393). It follows that there are places outside of and
beyond the academy where those of us who have been historically excluded
can engage our historical modes of survival to find our breath, where we
can be and do whatever, and where we can just *be*.

Being is not meant in the sense that we are haphazardly together: it means
to just *be* in the sense that we as Imani Perry writes, are in the process of
deliberately curating a world where we can "care for our souls and do so
with discernment" (2018, 228). Perry's discussion of curation deserves quoting at length:

> To curate in our lives is to practice deliberateness about what kinds
> of work we allow to come into view, not as blinders on the world, but
> as a means of cultivating the just imagination, as a means of devel-
> oping an active relationship to the world around us with purpose
> (and not simply for pleasure). At the same time, we must critically
> reflect upon our desires (knowing how our yearnings are shaped by
> ideology of the society, knowing that they very well may be doing
> damage to ourselves and others), with deliberate cognizance of the
> repeated fact that, notwithstanding the amazing range of representa-
> tions that we find in the simulacra, the status of nonpersonhood is

common and mundane both locally and globally. Therefore, curation of this sort must include the ugliness as well as the beauty, suffering as well as joy, and allowing – even requiring – of ourselves that we be moved by them all. Resisting the logic that creates nonpersonhood status requires something much more than people watching. Part of the work of curating has to be recognizing that markets can so often seduce us into mere watching instead of witnessing. Neoliberalism moves fast; it integrates the appearance of counter-hegemony at lightning speed. Evidence of something more or different happening than mere consumption can be found only in our deeds and our relations (2018, 228–229).

At this point, I wish to devote time to some particular terms: I have been repeating the words *whatever* and *elsewhere,* and this usage is purposeful. Here I shall further explain what I mean when I invoke these in the context of my text. I draw on Black studies scholar Rinaldo Walcott's and Black feminist Carole Boyce Davies' work for these terms. Walcott writes about a whatever of Black Studies that

is too disturbing to tolerate for some. It is not a simple and uncomplicated whatever. It is a 'whatever' that refuses the regulating and restricting confines of 'the nation-thing' discourse in the context of the study of black peoples. This is particularly important if we read blackness as a sign within (post)modernity that has always been beyond 'the nation-thing.' Denied citizenship within specific nations, black diasporic peoples have always had to articulate relations between place/nation and transnational identifications which position them in ambiguous relation to both the structures of nation and their narratives and to black people in other places (2003, 117).

Whatever as a concept therefore is expansive, leaving open all kinds of possibilities, all types of explorations, all types of configurations which are created outside and not restricted to the nation-thing of which the academy is a part and complicit in propping up. When I imagine what whatever looks like, I think of the unbothered shoulder shrug and knowing facial expressions, eyebrows slighted raised as one exclaims, "And so what? Its whatever!" and deliberately walks away from the nation/university-thing with a slow confident stride to the elsewhere that does not require us to be publicly complicit in our own oppression. It is more than a speaking back, it is a speaking beyond ambiguity; it is a form of reading and beyond shade.

Black feminist Carole Boyce Davies theorizes elsewhere similarly. In her book *Black Women, Writing and Identity: Migrations of the Subject,* Davies points out that Black women's writing "exist more in the realm of the "elsewhere" of diasporic imaginings than the precisely locatable.

Much of it is therefore oriented to articulating presences and histories across a variety of boundaries imposed by colonizers, but also by the men, the elders and other authorized figures in their various societies" (1994, 88). Avtar Brah also writes that "the concept of diaspora offers a critique of discourses of fixed origins while taking account of a homing desire, as distinct from a desire for a 'homeland'. This distinction is important, not least because not all diasporas sustain an ideology of 'return'" (1996, 16). It is worth noting that in her elsewhere search, paying close attention to that homing desire, Davies has rediscovered for us Claudia Jones, the Trinidadian-born Black feminist organizer, revolutionary, daughter of the diaspora, founder of the Notting Hill carnival. In her documentation of Jones' life, Davies writes, "it is important to give ... recognition to the kind of intellectual work produced organically outside of the academy [elsewhere] and accord that work the same weight and space one gives to academic production. ... one has to undo the narrow equivalence of intellectual work with the academy" (2007, 10). She continues "[i]t is in this context that we account for the activist-intellectual ... someone who was solidly located outside of any academic context but whose entire production of ideas rivaled many of those produced in the universities at the same time. This intellectual contribution is particularly important since black communities did not have the kind of access to academic institutions that they do following the civil rights era" (2007, 10). By locating Jones, Davies demonstrates the importance of looking for elsewhere spaces for those of us whose bodies sit uneasily within the constructs of the nation-thing. When I say Davies "found" and "located" her, I mean that Claudia Jones had been all but completely expunged from Black history. What seemed to be left of Jones in the archive were files that criminalized her, but Davies' search in the elsewhere places to uncover the activist and intellectual legacy of Jones demonstrates how Jones was able to imagine and create a world where diasporic Caribbean people then and now can satisfy their homing desire.

Thinking of elsewhere in the way that Davies writes about Jones invokes Avtar Brah's work on diaspora where she writes of the "contradictions of and between diasporic location and dislocation" (1996, 16). Brah refers to "...'diaspora space' ... [which] is 'inhabited' not only by diasporic subjects but equally by those who are constructed and represented as 'indigenous'. As such, the concept of *diaspora space* foregrounds the entanglements of genealogies of dispersion with those of 'staying put'" (1996, 16). For Brah, the story of the diaspora when contrasted with the nation-state, has subversive reverberations, as the diaspora has "little regard for national boundaries" (Hanchard 1990, 40). Davies also makes it clear how these diaspora spaces are constantly overlapping for people "displaced by global economic processes, who must constantly reconcile themselves to existing emotionally and physically in different spaces, may enter ... a diaspora, a space that resists centering even as it identifies longing, homelands, and a myth of origin. Still, there are those who remain outside a diaspora or who live in intersecting or overlapping diasporas" (2007, 21). By exploring the range, the experiences that are part of the diaspora, we can demonstrate the possibilities of elsewheres as "new geographical location[s] ... [and] different strategic location[s]" (Davies 2007, 120), which can stand in for the fullness, creativity and complexity of imaging and mapping possibilities across the nation/university-thing. As bell hooks states "'[t]he politics of location' necessarily calls those of us who would participate in the formation of counter-hegemonic practice to identify the spaces where we begin the process of revision" (1990, 145). This process of revision must be read alongside McKittrick's assertion that "one way to contend with unjust and uneven human/in human categorizations is to think about, and perhaps employ, the alternative geographic formulations that subaltern communities advance. ...remappings provided by black diaspora populations can incite new, or different, and perhaps more just, geographic stories" (2006, xix).

In *Demonic Grounds*, mentioned earlier in this book, Katherine McKittrick writes about what she calls Black geographies as different from transparent space. While transparent geographies define where some bodies naturally belong, Black geographies or unmapped knowledges expose the limitations of transparent space to speak back to colonialist grounded geographies or "the idea that space 'just is' and the illusion that the eternal world is readily knowable and not in need of evaluation, and that what we see is true" (2006, xv). She writes,

> [i]f *who* we see is tied up with *where* we see through truthful, common-sensical narratives, then the placement of subaltern bodies deceptively hardens spatial binaries, in turn suggesting that some bodies belong, some bodies do not belong, and some bodies are out of place. For black women, then, geographic domination is worked out through reading and managing their specific racial-sexual bodies. This management effectively, but not completely, displaces black geographic knowledge by assuming that black femininity is altogether knowable, unknowing, and expendable: she is seemingly in place by being out of place (xv).

If as McKittrick writes, traditional geographic arrangements both shape Black bodies and are challenged by them, then it means that these bodies, despite traditional ideologies and discourse that erases and distorts their belonging to a particular space (and time), are in fact geographic themselves (2006, xiii), in the sense that they hold within them the potential to reorient space not through management but through Lorde's conceptualizing of "chaos," as I will discuss in more detail in my concluding chapter.

McKittrick's theorizing of Black geographies engages with Édouard Glissant's concept of opacity where Glissant writes, "[t]he thought of opacity distracts me from absolute truths whose guardian I might believe myself to be. Far from cornering me within futility and inactivity, by making me sensitive to the limits of every method, it relativizes every possibility of every action within me" (1997, 192). By way of the example of Harriet Jacobs/ Linda Brent, author of *Incidents in the Life of a Slave Girl*, McKittrick explicated how these geographies are the last place they [white colonialist] thought of. Reframing geographies in this way is the antithesis of a black bodily management project but, according to McKittrick, requires that our energies extend to "the practice of mapping, exploring, and seeing, and social relations in and across space" (2006, ix), that we imagine and work toward finding our elsewheres.

The university that exists in the afterlives of slavery continues to obscure the knowledges and spaces of the marginalized and minoritized through "the protocols of intellectual disciplines" (Hartman 2008, 10). Therefore locating these knowledges and spaces means that we need to engage different or creative types of reading practices searching in the last place they/ we might have thought to look. In the university, we often seek to find and excavate from history by examining data, or texts, in the archives, but as Hartman writes in "Venus in Two Acts," these spaces are 'little more than a register of ... encounter[s] with power" (2008, 2) were we show up as non-persons, distorted, dismembered, dehumanized, artifact, as ditto, ditto, ditto.

Looking elsewhere to the nontransparent means following Hartman, we must engage a practice of critical fabulation, to read for "how might it be possible to generate a different set of descriptions" (2008, 7) acknowledging unheard voices and invisible lives. Locating these spaces within the epistemic violence of the transparent space of knowledge production poses particular challenges because of the ways in which blackness has been adversely shaped by traditional epistemic and geographic rules. However, in locating them, "...we see that black women's geographies are lived, possible, and imaginable. Black women's geographies open up a meaningful way to approach both the power and possibilities of geographic inquiry" (McKittrick 2006, ix) and move us beyond the space of the academy.

Locating elsewheres provides diasporic analytical openings to advance creative acts, such that to locate these elsewhere places, which are "fragmented, subjective, connective, invisible, visible, acknowledged, and conspicuously positioned" (McKittrick 2006, 7), requires a different type of looking, listening, seeing, reading. Returning to Walcott's "nation-thing," we understand the elsewhere place as outside its purview, a geography that is "both local and beyond local ... [that] allows for uncovering and returning these other spaces, places and sites of belonging ... [that] are ephemeral imaginary spaces ... [which] does not allow for any too-easy boundaries of restriction ... [and for] possibilities for transformation" (2003, 119). In thinking about how elsewhere spaces are invisiblized by traditional forms of geographical arrangements and the mapping of knowledge and bodies within the academic space, I am reminded, for example, of Nanny of the Maroons who hid and protected her people in the mountains of Jamaica and, how as legend has it, was a healer and a warrior. I first learned the bigger picture of Nanny's story, and how her unwillingness and indifference to European norms gained her such notoriety through the work of Black feminist Andrea Shaw, who engaged in a deep diasporic reading in her 2006 book, *The Embodiment of Disobedience: Fat Black Women's Unruly Political Bodies.* Although her history is rich and exciting, I had to really search to find out more about Nanny beyond how she was mythologized or rather dehumanized by Europeans, stripped down to just her body parts, in particular her buttocks (grotesquely, they wrote that she would shoot bullets therefrom).

As I think about Nanny of the Maroons, I also think of others such as Harriet Tubman, who inhabited physically opaque/hidden spaces. As Manu Vimalassery writes, I also imagine Tubman must have conveyed the significance of meaning

> through changes in timbre, her choice of ornamentation, or by altering scales... the rhythm of her breath, or her pause ... Each of these attributes is described in the archive of her life and actions, but each of these, we can only imagine. Her distance from literacy is an immersion in a culture of orality and living embodiment, and the result, for us, is not so much a production of ignorance, as realms of knowledge that are irrecoverable through familiar empirical methods (2016, n.p.).

To find these nuances that hold in them the possibility of making life anew, then we must read differently, listen harder, see more intently, look elsewhere – in the last place they thought of but "which have always existed before our very eyes" (McKittrick 2006, 8). Vimalassery observes that "[w]hile Tubman's immersion in orality poses challenges for methods suited to the colonial and slaveholding archive, her own reading practices might point us to different ways of inhabiting place, different terms of relationship that chafe against the assumptions of the racial state" (Vimalassery 2016, n.p.), and the university.

Black women and by extension Black feminisms have always engaged alternative ways of knowing as valid forms of knowledge and knowledge production while also acknowledging the unknowable. For example, we are intimately familiar with the ways in which encounters with power have erased the lives of many of our ancestors from the archives, as is the case with Venus, an African girl who like many others, kidnapped from her home was murdered onboard a slave ship bound for the United

States. All we know about Venus is from a cursory mention of her in court documents. McKittrick (2006) implores Black feminists to think about and acknowledge the *where* of race and Black feminisms, the place where Black feminists move from and which "undermines the norms of scholarly authority and mastery upon which the university is based" (Hong 2008, 106), to demonstrate that the institution's definitions and reproduction of knowledge are limited and limiting. As bell hooks writes, when we move from the elsewhere of Black feminist knowledge, we acknowledge the partiality of university mastery: "[w]e come from a long line of ancestors who knew how to heal the wounded black psyche when it was assaulted by white-supremacist beliefs. Those powerful survival strategies have been handed down from generation to generation. They exist. And though a working public knowledge of them has been suppressed, we can bring this old knowledge out of dusty attics, closets of the mind where we have learned to hide our ghosts away, and relearn useful habits of thinking and being" (2005, 61).

The elsewhere of Black feminisms then can be all at once a physical, cultural, ancestral, imaginative, theoretical, and affective space. Our insisting on Black feminist politics as a location exposes how speaking and moving from this place has the potential to disrupt academic metanarratives and is a place from where we can begin to locate and thus make demands for "all sorts of political positions and connections without distorting the theoretical possibilities and material realities of spaces unheard, silenced, and erased" (McKittrick 2006, 56). Because our knowledges have been subjected to erasure and made invisible, Black feminists are committed to handling it and us both carefully and with care, such that according to Jennifer Nash, Black feminists are grounded in and have a "long-standing commitment to love-politics, to ethics of mutual vulnerability and witnessing" (2019, 130). We can use the where of Black feminisms as a place to think about how we can "work within, think within, practice poetics of peace and equity from within the [academic] space while also remaining critically aware of how the space would attempt to exploit the very projects against the university in the service of the flourishing of the neoliberal logics of the university... This is the challenge. It is also the opportunity. It is a gap to which we must be mindful, a space and break of possibility. It is in that space that the flourishing of radical thought and mood and movement can occur" (Crawley 2018, 8). Like the enslaved woman Delicia Patterson who engaged in a lifetime of refusals to engage in bodily management projects that would make her legible to fit the dehumanizing constructs of white slavers – most notably her public refusal on the auction block to be sold to Old Judge Miller, one of the wealthiest but meanest slave owners (McKittrick 2006, 83) – when we move from the elsewhere of Black feminisms rather than from the transparent transactional space of the university, we too are able to dispense with bodily management as pragmatism and be confident in our right to radical refusal.

I, like many other Black, and decolonial and transnational feminists including M. Jacqui Alexander and Chandra Talpade Mohanty see the university "through a series of geographical, political, and intellectual dislocations" (2012, xiii). I declare, in agreement with Felly Nkweto Simmonds, that though I am an academic, the body that I live, feel, and think in exists fundamentally in conflict with what it has traditionally meant to be an academic (1999, 228). Like other Black feminists, I get how the corporeal body can be deadened within an institution that seeks to awaken the mind. I "recognize the corporeality of our bodies rather than abstracting and quantifying them into 'the body,' a body no one lives or breathes [or feels] within" (Chinn 2000, 169).

The hereness of traditional geographic academic space and spatialization requires the ongoing placelessness and displacement of the marginalized and minoritized, such that we are compelled to enact a level of pragmatism which puts us in our assigned place (McKittrick 2006, 8-9). Seeking and moving toward an elsewhere then is inherently subversive in its refusals to accept placelessness. As Nash puts it "the experience of being black feminist engenders certain kinds of feelings in its practitioner, feelings of fatigue, of sadness, of anger. ... to claim black feminism as one's academic home is an experience that has an affective charge" (2019, 29). Black feminists and by extension Black feminisms then, gets to the affective, so to engage in feelings of determination and yet playfulness. As I see it, affectivity is vital to worldmaking. Instead of Black feminisms being "a coffin and grave" (Stallings 2011, 138) they are a vibrant landscape of elsewhere. Embracing the whatever of Black feminist affectivity means that we remain in conflict with the way space is socially constructed, and us within it, such that we in the words of Lugones do not end up "a serious human being, someone with no multi-dimensionality, with no fun in life, someone who has had the fun constructed out of her. I am seriously scared of getting stuck in a 'world' that constructs me that way. A 'world' that I have no escape from and in which I cannot be playful" (1990, 399). Experiencing Black feminisms as a location and an elsewhere, where we are able to engage in all types of critical affective reading practices, is essential "[f]or our feelings are the sanctuaries and spawning grounds for the most radical and daring of our ideas" (Byrd, Cole and Guy-Sheftall 2009, 186)

Inhabiting the elsewhere of Black feminisms also means that we first and foremost evaluate honestly the things for which we yearn. As Black feminist Angela Davis puts it in her 2016 text *Freedom is a Constant Struggle: Ferguson, Palestine, and the Foundations of a Movement,*

> Everyone is familiar with the slogan "The personal is political" – not only that what we experience on a personal level has profound political implications, but that our interior lives, our emotional lives are very

much informed by ideology. We ourselves often do the work of the state in and through our interior lives. What we often assume belongs most intimately to ourselves and to our emotional life has been produced elsewhere and has been recruited to do the work of racism and repression. ... This means we have to examine various dimensions of our lives – from social relations, political contexts – but also our interior lives. It's interesting that in this era of global capitalism the corporations have learned how to do that: the corporations have learned how to access aspects of our lives that cause us to often express our innermost dreams in terms of capitalist commodities, So we have internalized exchange value in ways that would have been entirely unimaginable to the authors of *Capital* (142).

Inhabiting the elsewhere of Black feminisms means being attentive to the ways in which the things we desire most and the avenues through which we try to achieve those things might be contrary to our own liberation. Inhabiting Black feminisms as an elsewhere place allows us to be discerning through the ways in which we listen closely and feel reverberations in the way that Tubman intended, and respond in kind to them as if our very life depended on it. Inhabiting Black feminisms as an elsewhere place means, as Perry writes, "doing, as diachronic poesis of living in politics that is ever changing, uncertain, and vexed but that also, we hope, will bring us closer to freeing us all" (2018, 253).

The elsewhere requires we have imagination and see the possibilities in creative and nonapparent ways of being. It can be described as McKittrick writes, "rhizomorphic, a piece of the way, diasporic, blues terrains, spiritual" (2006, 7). Building upon Glissant's poetics of landscape, McKittrick writes that these Black feminist elsewhere spaces contain "theories, poems, historical narratives that disclose black women's spaces and places" (2006, 7). What we know of these poems and narratives is that in many instances, they offer both literal and figurative lifegiving, as without them most of what we would know would be about Black people's encounters with white colonial power through archival reports created or adapted mostly by the colonizers. The poems and narratives of elsewhere spaces are sometimes all we have to make the lives of our ancestors known to us, but this in turn can still work to breathe fresh air into our bodies to hold on, to move us and get us through. These narratives and poems must not be the last place we think of or look, rather we must look to them always and often for aspiration.

Aspiration and the possibilities of the erotic and the poetic

Black feminist Imani Perry, in her 2018 text *Vexy Thing: On Gender and Liberation*, writes about maps as sites of contestation, as places for alternative imaginings and naming of relations and representations. If the purpose of maps is to draw our attention to one set of things rather than another, in meta-phorical form of symbol or color or letters, to do something, then that something always has values and ethics attached to it, whether visible or invisible. ... Remapping terrain, history, and the body have been important means of challenging and even dismantling neocolonial cartographies of human value. Further, alternative maps guide us to think deeply about and confront the ethics of what and who garners our attentions and commitments" (2018, 179).

Looking toward Black feminisms as an elsewhere place requires (re)map-ping in the way that Perry suggests, in order to make perceptible to us what inhabiting this elsewhere can make possible. That the elsewhere of Black feminisms presents as a theoretical and political project that is lived and felt means that its spatio-temporal locations are not always visible within transparent space. As such, we have to ask ourselves, how does the habitation of Black feminist elsewheres work to respatialize seemingly natural but socially produced spaces and places of subjugation into sites where we can assert our human value (McKittrickk 2006, xviii-xix)? One might argue that remapping as a means of asserting alternative relations to space, narratives, historical routes/roots, and representations has to start with addressing the affective. If remapping, as according to Perry, is about reassessing human value, then a critical engagement with how devaluation feels is important to sensing what types of new terrain we can imagine, and to our commitment to finding spaces where our humanity is celebrated and where we can actively celebrate, that is share joy in these created spaces. This remapping cannot be guided solely based on what we know cerebrally but must also be centered in the ways of *just knowing* how we feel in our interior. As Black feminists Joan Morgan (2000) and more recently Tamura Lomax remind us, our interior spaces are sacred. Lomax, for example, writes of the interior as a "space beyond or prior to knowledge of signification as well as intervening space between rep-resentation and the interior, despite how forceful or cross-pollinating the projection" (2018, 137).

As Jennifer C. Nash puts it, our interior is "always transcending attempts to limit" (2019, 5). It is the invisible key/symbol that guides us as our cartographical selves. Being in tune with our interior means that our yearning takes on a different significance, that the things and places and people for which we yearn are not easily coopted by external forces, nor are we misguided by the exterior, neoliberal measurements of suc-cess. Moving from the interior, whatever yearnings we have to belong or to achieve, cannot be confined "to membership of citizenship in commu-nity, political movement, nation, group, or belonging to a family, however constituted, although important" (Alexander 2005, 281-282); rather our yearning is for a place, an elsewhere where we matter, where our humanity is valued, where our "people and community matter and where spiritual and emotional growth matter" (hooks 2005, xxi). When we recognize our yearnings originate in/from that elsewhere of the interior and from "the deep knowing that we are in fact interdependent – neither separate nor autonomous," and we come to understand that "[a]s human beings, we have a sacred connection to one another, and ... enforced separations wreak havoc on our Souls" (Alexander 2005, 282) we are able to keep alive and breathe new air into the creative (re)mapping work that keeps us alive. Speaking and doing from the location of Black feminisms, the elsewhere ultimately means that we don't have to speak as or from a place of difference as the university's proof of diversity, that we can speak from our interior elsewhere. We are not simply provoked to the reaction by an exterior or motivated by how closely we can locate ourselves in proximity to neoliberal notions of success and power. We engage instead in what Kevin Quashie calls "an ethic of quiet, the sense that the interior can inform a way of being in the world that is not consumed by publicness but that is expressive and dynamic nonetheless" (2012, 52).

The elsewhere of Black feminisms is quiet, not in the sense that it is not or cannot be loud or expressive, because as Davies, quoting Charles Nero (2005), writes, "'capping, loud-talking, the dozens, reading, going off, talking smart, sounding, joining (jonesing), dropping lugs, snapping, woofing, styling out, and calling out of one's name' (p. 230) ... [are] [a]ll ways of "reading" other's behaviors and practices ... often directed at dismantling dominant or pretentious discourses" (1994, 42-43). These gestures of speaking back to dehumanizing Eurocentric discourses about the Black body are creative, and as M. NourbeSe Philip describes, they involve "the whole body... gestures, arms akimbo" (McKittrick 2015, 155). These expressions originate from the creative insides of Black people, an elsewhere that is quiet, one that "read[s] ... expos[ing] life that is not already determined by narratives of the social world" (Quashie 2012, 8). To further explain what I mean by "quiet," I draw from Quashie, who writes, "[i]n humanity, quiet is inevitable, essential. It is a simple, beautiful part of what it means to be alive. ... it requires a shift in how we read, what we look for, and what we expect, even what we remain open to. It requires paying attention in a different way" (2012, 6). Paying attention in a different way may resemble trusting our guts, looking beyond the obvious to observe intently situations and movements, watching, looking, listening in ways that can produce all types of creative expressions as maps back to our ancestors and through them find our own way. We can build on what we know, feel, imagine, find the ancestors, to create and express the types of knowledge mentioned above, knowledge that may be unrecognizable to others as knowledge but that are rich and beautiful ways that allow us to be our own human selves whether or not our being is affirmed or legitimated by those with power. Paying attention in a different way is the promise of a

Black feminist Shad(e)y theoretics, which after a good read can deftly expose the ugliness of the transparent for what it is. Paying attention in a different way opens up all types of refusals that are not simply readable as resistance and do not require permission, but which allow us to keep our breath and to breathe life.

A Black feminist elsewhere does what Christina Sharpe refers to as Wake work, in that it is "aspiration, that keeping breath in the Black body" (2016, 109). Many of us, when we think of what the term woke/wokeness mean, think about that we need to do, specifically as Black folks, to be and remain aware. But being aware and staying aware encompass only a part of doing wake work. Moving from that place of awareness is also important not merely as a performative act but as part of being in relation with a community that is literally having its breath choked and kneed out of it, as Black bodies are saying with their last gasps for air, "I can't breathe." While woke is past tense, wake situates us squarely in the present, the doing, the trying to keep breath in living Black bodies and to keep breathing life into the memories of those made breathless in the afterlives of slavery, while working toward securing all types of air space for future generations. Wake work is paying attention in a different way, it is to engage in (re)mapping. As Sharpe writes,

> Living ... in the wake of slavery, in spaces where we were never meant to survive, or have been punished for surviving and for daring to claim or make spaces for something like freedom, we yet reimagine and transform spaces for and practices of an ethics of care (as in repair, maintenance, attention), an ethics of seeing, and of *being* in the wake as consciousness; as a way of remembering and observance that started with the door of no return, continued in the hold of the ship and on the shore. ... This is an account counter to the violence of abstraction ... An account of *care* as shared risk between and among the Black trans*asterisked (2016, 130–131).

According to McKittrick, this account is "recognizing both 'the where' of alterity *and* the geographical imperatives in the struggle for social justice" (2006, xix). Black feminists then are creating elsewhere, moving from the where of Black feminism, by pulling from the interior, the ancestral, and the communal.

As I work through this concept of the Black feminist Shad(e)y theoretics, I am reminded again of the work of one of the "shadiest" Black feminist that has ever lived, Audre Lorde. Lorde had no qualms throwing shade and did so multiple times, most famously in her letter to Mary Daly (1984) calling her in, and then out, on her book *Gyn/Ecology*'s exclusion and disfiguring of Black women's goddess heritage and sisterhood. Lorde's work on the power of the erotic, which is deeply instructive to mine, has been figuratively giving Black women life (even in death): she has been breathing life into our bodies, she is a source of aspiration. Thinking about Lorde's work in this way, I am drawn to thinking about the erotic and her own poetry, as well as poetics in general as the type of whatever that searching for and finding the elsewhere produces. Édouard Glissant writes that poetics are crucial to the elsewhere: "[t]o consecrate the union between elsewhere and possibility, the poet ...kept a resolute distance from any Here conferred in advance (not willfully mediated)" (1997, 37-38). The here of the academy we inhabit has been thrusted upon us and configures who we are scientifically, ideologically, philosophically, and politically in advance of our appearance/inclusion, first as nonhuman and nonpersons, and now as difference and diversity. This cannot be our avenue, passageway, route to something or somewhere better; it is a dead-end! The here of the academy forecloses the possibilities of the elsewhere and whatever inventiveness our imagination can conjure there.

My thoughts here turn to Katherine McKittrick, who writes, invoking Sylvia Wynter: "human life is marked by a racial economy of knowledge that conceals – but does not necessarily expunge – relational possibilities and the New World views of those who construct a reality that is produced outside, or pushing against, the laws of captivity. It follows, according to Wynter, that we would do well to reanimate and thus more fully realize the co-relational poetics-aesthetics of our scientific selves" (2015, 8). Through McKittrick's words, my thoughts also connect with Hartman's writing regarding the impossibility of resuscitating Venus from the archives. Hartman writes, "The necessity of trying to represent what we cannot, rather than leading to pessimism or despair must be embraced as the impossibility that conditions our knowledge of the past and animates our desire for a liberated future" (2008, 13). Constantly crashing into these impossibilities can and do stir up that chaos within us; that lens, Lorde writes, through which we can scrutinize all aspects of our existence. It spurs us to (re)map, not using those cartographical traditions used to create here space, but those that can help us to find that union, according to Glissant between elsewhere and possibility. Sharpe writes of Black feminist Dionne Brand's poetry, specifically the poem entitled "Ruttier for the Marooned in the Diaspora," that Brand "maps the desire to say more than what is allowed by an archive that turns Black bodies into fungible flesh and deposits them there, betrayed... her offering to guide us to how to love in the wake... is a guide of indiscipline and lawlessness; a map of disinheritance and inhabitation" (2016, 132). In this reflection on Brand, Sharpe really gets at the importance of poetics for breathing life into us through refusal, being indisciplined and lawless rather than through pragmatism. We must disinherit what is not our birthright, refuse the racist, sexist heavy weight of our scientifically/pseudoscientifically tallied bodies, for it is meant to asphyxiate us.

The concept of relations or relationality is also explored by Glissant in *Poetics of Relation*. There he writes "Relation is not an absolute" (1997, 35). I interpret this to mean that relations change over time and space, and so how we see ourselves as well as our current and future relations with and within the university is up for grabs. Thinking through and working toward a decolonial relationship with the university in a future elsewhere which would hold space for us to breathe life into physical and theoretical bodies, bodies of knowledge which were once deadened by and are currently being killed by the modern imperial university would also include Kim Tallbear's observation that

> "we work ourselves into a web of relations (I am thinking in terms of space and not a time concept now). ...Can such disaggregation help us decolonize the ways in which we engage other bodies intimately – whether those are human bodies, bodies of water or land, the bodies of other living beings, and the vitality of our ancestors and other beings no longer or not yet embodied? By focusing on actual states of relations – on being in good relation-with, making kin – and with less monitoring and regulation of categories, might that spur more just interactions? (2018, 161)

What kind of future elsewheres would be possible if we could accomplish this type of intimacy? What type of (re)mapping does this intimacy make possible by allowing us to imagine and explore new terrain which does not require the knowledge of traditional archives, maps, charts, official records or scientific figures, but rather, as McKittrick writes, "uses voice-language, a poetic-politics, and conceptualizes ... surroundings as 'uncharted,' and inextricably connected to ... selfhood and a local community history" (2006, xxii)? How would we be able to see "the world from an interhuman (rather than partial) perspective" (McKittrick 2006, 144), where both "[d]weller and pilgrim live this same exile" (Glissant 1997, 41)?

When I think of the possibility of intimacy, poetics, and mapping the uncharted, it brings to mind Alice Walker and the poetic, genealogical, cartographical work she performs in her 1983 text, *In Search of Our Mothers' Gardens: Womanist Prose*. Alice Walker's definition of womanism pulls from the elsewhere archive of Black women's gardening and quilting, archives overlooked because they were in the last place they thought of. Walker provides a map which according to Nash, "places self-love at the center of black feminist subjectivity." Further pursuing her politics of care, Nash writes, "the Combahee River Collective's statement that its political work emerged from 'a healthy love for ourselves, our sisters, and our community,'" and demonstrated how "black feminists have long emphasized the importance of love as a form of collectivity, as way of feeling, and a practice of ordering the self" (2019, 115). When Alice Walker coins the term womanist/womanism, she invokes the poetic-politics McKittrick mentions, that is, she values and uses the elsewhere knowledge of Black women's creative lives and everyday lived experiences to write, "*Womanist is to feminist* as purple is to lavender" (1983, xii). This one sentence conveys so much meaning about the richness and fullness of Black women's lived experiences. It contains within it the power of paying attention in a different way, such that we can see and pull from our own ancestral archives to map our own future. Walker's declaration that a womanist is "outrageous, audacious, courageous or willful ... Wanting to know more and in greater depth than is considered "good" for one. ...Committed to survival and wholeness of entire people ... Loves music. Loves dance. Loves the moon. Loves the Spirit. Loves love and food and roundness. ... Loves the Folk. Loves herself. Regardless," (1983, xi-xii) is the type of indiscipline Brand evokes in "Ruttier," the type of indiscipline enacted by Delicia Patterson, which shows up in the reading, snapping, talking smart, shading that is essential to dismantling dominant discourse. The sentiment that "*Womanist is to feminist* as purple is to lavender" is shady as fuck! A Black feminist Shad(e)y theoretics is not interested in a body management project, it is not a pragmatic project – it is unruly, it is a refusal, and in that refusal, it is breath.

According to Davies, Black women's writing is "a series of boundary crossings ... not ... fixed" (1994, 4) such that "[t]he ways in which Black women/women of color theorize themselves often remains outside of the boundaries of the academic context, or 'elsewhere.'" (18). In the concluding chapter to this text, I posit that the poetic knowledge (the whatever) that comes from being in relation within the elsewhere space has transformative, creolizing potential and is "born of a reality of relation [that] prophesy or illuminate ... divert ... or conversely gain strength within ... mak[ing] ...[a] sort of music" out of the conditions of being in relationship with each other in the wake (Glissant 1997, 93-94). Christina Sharpe, mentioned earlier, writes about being in the hold. In a similar vein, Harney and Moten identify the hold as a place, a where of Blackness; but they also note: "while certain abilities – to connect, to translate, to adapt, to travel – were forged in the experiment of hold, they were not the point" (2013, 97). Harney and Moten underscore the possibility of the hold as a terrible and terrifying place which at the same time gave us the dreadful gift of gathering "dispossessed feelings in common, to create a new feel [when] [p]reviously, except in these instances, feeling was mine or it was ours." In other words, it allowed us to feel how when "Black Shadow sings 'are you feelin' the feelin?' he is asking about something else. He is asking about a way of feeling through others feeling you" (97-98). As a Caribbean woman who has been to many a fete and heard the DJ shout, "How you feeling tonight?" to hype up the crowd, this makes intuitive sense – the energy that comes from being in relation together – intimate in the wake of the hold.

Like Nash, I use the terms intimate or intimacy because of the way it suggests "permeability between concepts and their 'origins,' and between bodies ...including the possibility of being done and undone through relationality" (2019, 107). I propose, therefore, that the subjective and individual affective produces a sense of and a sensing for a Black feminist elsewhere/whatever which exists as "something not necessarily corporal ... [but] engages many bodies at once, rather than (only) being contained as an emotion within a single body" (Chen 2012, 11). According to Mel Chen, "[a]ffect inheres in the capacity to affect and be affected. ...the entry of an exterior object not only influences ... further affectivity ... but 'emotions' that body: it lends it particular emotions or feelings as against others" (2012, 11-12). I connect Chen's concept with what Harney and Moten call Hapticality, or, "the capacity to feel through others, for others to feel through you, for you to feel them feeling you[;] this feel ... is not regulated, at least not successfully, by a state, a religion, a people, an empire, a piece of land, a totem. Though forced to touch and be touched, to sense and be sensed in that space of no space, though refused sentiment, history and home, we feel (for) each other" (2013, 98). This feeling is visceral, "raw and intimate" such that it allows us to witness each other, that is "to feel as well as to see me, to allow yourself to be moved by me toward a different self and a different relation to me, and potentially the world" (Perry 2018, 217). Being affected by each other in this way, we can be, make, and engage in a common cause that refuses those constructions of ourselves that justified putting us in the hold in the first place. Being affected by each other, feeling us, can produce the types of refusals that pragmatism makes unimaginable, a refusal that can, in the words of Jack Halberstam, "reshape desire [yearning], reorient hope, reimagine possibility and do so separate from the fantasies nestled into rights and respectability" (2013, 11–12).

We witness that a Black feminist elsewhere is also a space of recovery. bell hooks writes about how our oppressors benefit when we are depleted and "have nothing to give our own, when they have so taken from us our dignity and our humanness that we have nothing left, no 'homeplace' where we can recover ourselves" (1990, 43). It follows that sensing a Black feminist elsewhere also builds a sense of home/homeplace which (while it may not be a physical homeplace, and may be better defined by "its nowhereness and everywhereness" (Davies 1994, 103-104)) is still a space where we can be and rest. The concept of a Black feminist elsewhere as home must be juxtaposed with the ways in which we think about exile and the exilic, particularly in the sense that Kevin Quashie writes. According to Quashie, the exilic comes from a sense of freedom, vagary and wonder, the inner life (2012, 129). As such, it resonates with Davies when she states that "homelessness itself cannot be trivialized or essentialized into a flat, monolithic category ... [f]or some ... exile is a desired location out of which they can create" and be inspired (1994, 114). In effect, the desire for home and for exile both represent a desire for places (alternative geographies and relations) where we can find rest and inspiration through finding ourselves or finding our people. Recovery through the possibilities of wonder and wander, or being in relation with the familial and familiar, holds out a sense of the immense possibilities of Black feminist elsewheres, and this allows us to decipher clearly our own selves and our own needs. Like Achille Mbembe, I believe that "[w]e are called upon to see ourselves clearly, not as an act of secession from the rest of the humanity, but in relation to ourselves and to other selves with whom we share the universe. And the term 'other selves' is open ended enough to include, in this Age of the Anthropocene, all sorts of living species and objects, including the biosphere itself" (2015, 16–17). Other selves include all that we encounter as we wander.

Moving from these places of being in relation - these elsewhere affective spaces - in spite of and despite the university's hold, is powerful. We know that everyone feels, but in the space of the university, our feelings have been foreclosed in the name of survival (Quashie 2012, 66). Yet, we also know that there is power in feeling, the type of feeling which Lorde calls the erotic. She tells us that opening up to this "erotic knowledge empowers us, becomes a lens through which we scrutinize all aspects of our existence, forcing us to evaluate those aspects honestly in terms of their relative meaning within our lives. And this is a grave responsibility, projected from within each of us, not to settle for the convenient, the shoddy, the conventionally expected, nor the merely safe" (1984, 57). Conceiving of the elsewhere as an affective space means that we do not have to "privilege literal movement, the crossing of state borders or some other engagement of social or political institutions; [but engage] ... consideration of the mobility that is part of interiority, the inevitable human capacity to wander without ever taking a step" (Quashie 2012, 125–126). According to Davies, "the struggle over the signs of captivity is consistently expressed in rejection of current conditions, physical and emotional movement, assertiveness and a variety of migrations to 'elsewhere.'" (1994, 134). These varieties of movements, including emotional movement and affective wander, make no claim to scientific legitimacy or entitlement (Glissant 1997, 144), only to our multiple ways of being human.

The importance of the interior to finding our elsewhere can be seen in the way the interior allows us to untangle ourselves from the promise and politics of yearning to rely instead on the erotic. Quashie writes: "[b]ut all is not the world outside; the inside too is a life, one that is capable of feeling and knowing and pleasure. And the human heart, powered by the agency of one's inner life, is agile enough to negotiate the world. ... it runs against the expectations of the social world" (2012, 68). While yearning is susceptible to co-optation and manipulation, Lorde lets us know that the erotic is not so fragile, because engaging it means that we "live from within outward ... recogniz[ing] our deepest feelings" (1984, 58). Riffing off of Quashie again, I reflect similarly about the possibilities of the interiority of quiet, which he describes as "like the wander of water, the freedom in being lost or compassed by vagary. It is almost an exilic condition, this quiet that is the inner life of every human being" (2012, 129). He continues,

> [m]ost simply, interiority is a quality of being inward, a "metaphor" for "life and creativity beyond the public face of stereotype and limited imagination ... The interior is the inner reservoir of thoughts, feelings, desires, fears, ambitions that shape a human self; it is both a space of wild selffullness, a kind of self-indulgence, and "the locus at which self interrogation takes place" (Spillers, *Black, White, and in Color,* 383). Said another way, the interior is expansive, voluptuous, creative; impulsive and dangerous, it is not subject to one's control but instead has to be taken on its own terms. It is not to be confused with intentionality or consciousness, since it is something more chaotic than that, more akin to hunger, memory, forgetting, the edges of all the humanness one has. Despite its name, the interior is not unconnected to the world of things (the public or political or social world), nor is it an exact antonym for exterior. Instead, the interior shifts in regard to life's stimuli but it is neither resistant to nor overdetermined by the vagaries of the outer world. The interior has its own ineffable integrity and it is a stay against the social world (21).

M. Jacqui Alexander writes of Tubman mentioned above that she moved from her interior (2005). Because the interior can only be known through expressions (Quashie 2012, 22), we can only guess at Tubman's through what we can know of her behavior and habits. And yet we deeply know that her desire for freedom and to free her people, to take them from the place of bondage, must have come not from pragmatist yearnings, but originated instead from the chaos of the interior desire to create new places through mapping out freedom over and over again, as manifested in the Underground Railroad and the light of the North Star.

Tubman expressed her interiority through her words, her dress, her movement. As Manu Vimalassery writes, "While Tubman's immersion in orality poses challenges for methods suited to the colonial and slaveholding archive, her own reading practices might point us to different ways of inhabiting place, different terms of relationship that chafe against the assumptions of the racial state" (2016, n.p.). Considering the importance of modulations in Tubman's voice, Alexander in turn points out: "[m]odulations in voice ... are not solely speech – perhaps not about speech at all – but instead are about an opening that permits us to hear the muse, an indication of how memory works, how it comes to be animated. But whose memory, whose voice, and whose history" (2005, 16)? This point is important to keep in mind as we think about how the interior space of memory and remembering, according to Alexander, is an "antidote to alienation, separation, and the amnesia that domination produces" (2005, 14), particularly as most of what we know about Black lives has been passed on and remembered not by official accounts of history, the "bias grains" as Fuentes (2016) describes, or the "normative thrust of historiographical inquiry" (Myers 2018, 109) which is the archives. So, the where of Black feminisms also requires remembering as a journey, "where we must return to track our difficult opaque sources" (Glissant 1997, 73). The concepts of interiority, quiet, memory and remembering have the potential to represent the whatever of humanity, in that they "can support representations of blackness [and otherness] that are irreverent, messy complicated – representations that have greater human texture and specificity" (Quashie 2012, 23) where we are not beholden to non-human and non-person stereotypes, but can be whatever and whomever we like. When we start with and move from our erotic, chaotic interior selves we know to desire "possibility rather than achievement... a giving into, a falling into self" (Quashie 2012, 45).

Returning to Sarah Ahmed's writing quoted in the Introduction of this text, we then need to ask not only how did we arrive here, but how is our arrival "linked to other places, to an elsewhere that is not simply absent or present? We also need to consider how the *here-ness* of this encounter might affect *where we might yet be going*" (2000, 145). How does existing in the university in the ways that we do today, how does experiencing it in the ways that we do, literally and figurative take our breath away, make us search for breath in between tears, bursts of anger and frustration, and exclamations of joy? How can we engage our interior, erotic, relational selves to map where we might yet be going, where we want to be? Riffing off of McKittrick once more, I see that it is "[t]he not-quite spaces of black women [that] provide alternative paths through traditional geographies and take into account a political agenda concerned with racism-sexism, objectification, captivity and respatalizations ...Black women's knowable sense of place is often still found in the last place they thought of across the logic of white and patriarchal maps" (2006, 62). Black feminist elsewhere spaces are spaces where knowledge is produced in the space of refusal, in the last place they thought of. Riffing off of Patricia Hill Collins, these places "have been essential in developing a way of knowing about our Blackness beyond that of a racialized spectacle. [In] ... kitchens, hair salons, everyday conversations, musicians, poets and writers as key to understanding how processes of racialization can be understood in different forms" (Johnson and Joseph-Sallisbury 2018, 151). For we have always had a truer word to say about ourselves, we have always theorized, but theorization is done relationally as we feel for a sense of each other, "in forms quite different from the Western form of abstract logic ... in the stories we create, in riddles and proverbs, in the play with language" (Christian 1987, 52). How else Barbara Christian asks would be have survived, had we not listened for, played with, and felt the pleasure, that is, had we not celebrated our ways of producing knowledge "both sensual and abstract, both beautiful and communicative" (52). We must feel this, we must imagine this, imagining it is bound up in our project of liberation (Hughes 2018, 165).

This liberation project is one which subverts "the [Euro]American talent for transforming bodies into things (information, statistics, evidence, databases)" (Chinn 2000, 171). It is creative, geographical worldmaking; a whatever project which allows us the freedom to be undisciplined, such that we may dance and sing only when and because we feel it, when we "feelin the feelin." This project attends "to the promise of science ... without reifying a biocentric ... worldview ... [by] creative narratives [that] point to the neurological, flesh, blood, and bones of humanness as these biologics are entwined with a racially structured discourse of condemnation" (McKittrick 2015, 155). I relate this directly to what McKittrick calls "the science of the word." By the science of the word, McKittrick means the ways in which we are able to describe our multiple ways of being human through the poetics of language that "emphasizes how we symbolically attend to, and *feel*, our genetic and biological world. The science of the word does not *describe* our surroundings, our flesh and bones, and our experiences; rather, it emphasizes how the emergence of the human was accompanied by, rather than preceded by, the word ... The science of the word, then, points to creative labor as recoding science through representational and biological feelings" (2015, 154-155). The science of the word allows us to engage in and explore creative, collaborative intellectual elsewheres which have been deemed unscientific, and incomprehensible (McKittrick 2015, 154–155) – Alice Walker's mother's garden, for example.

What we know about the hereness of inhabiting the transparent space of the university is that for Black and Brown people, women and others, left out of the enlightenment knowledge project, it often does not feel good. Like Clare, we feel like we are merely visiting, that it is not our homeplace, neither is it our place of wonder/wander. When we collectively start with the question, "'why doesn't this feel good?' ... this naturalization of misery, the belief that intellectual work requires alienation and immobility and that the ensuing pain and nausea is a kind of badge of honor, a kind of stripe you can apply to your academic robe or something. Enjoyment is suspect, untrustworthy, a mark of illegitimate privilege or of some kind of sissified refusal to look squarely into the fucked-up face of things which is, evidently, only something you can do in isolation"

(Halberstam 2013, 117–118). To feel the university in this way and be nauseated by it, particularly in the time of a pandemic when we are literally not feeling well, when we are already stressed out, anxious, tried, overworked, and some of us are literally fighting to breathe, as the university continues on, collecting more data to quantify and translate our bodies into productive units, to tell a pragmatic story of how it and we survived, we long for another time and space post-pandemic. But we cannot be so shortsighted. It never felt good, it does not feel good now, and it will only continue to feel worse.

The where of Back feminisms is aspiration(al), it can allow us to find our "Soul values, which ... [come] from deep within ... Unlike appraisal and seal values ... outsiders ... [do] not bargain for them" (Berry 2017, 61). If we consider the where of Black feminisms as we read the university, then the ultimate shade is to point out that *where* as a space(s) that can be inhabited by all who are invested, regardless of gender or race. The place of Black feminisms then is a decolonial place of opacity which can be found by those willing to look hard enough to find it. McKittrick tells us,

> The combination of diverse theories, literatures, and material geographies works to displace 'disciplinary' motives and demonstrate that the varying places of black women are connected to multiple material and textual landscapes and ways of knowing. These discussions are also about geographic stories. Places and spaces of blackness and black femininity are employed to uncover otherwise concealed or expendable human geographies. Because these geographic stories are predicated on struggle, and examine the interplay between geographies of domination and black women's geographies, they are not conclusive or finished (2006, xxxi).

Here we see the expansiveness of Black feminisms which Jennifer C. Nash writes about, a wide-ranging geography that is about Black women's humanity and visions for the future (2019, 5), one which Glissant says "clamor[s] for the right to opacity for everyone" (1997, 194). It is an elsewhere place that "is neither complete nor fully intelligible" (McKittrick 2006, 2). In Chapter 3 of this text, I mention briefly, using the work of M. NourbeSe Philip, how the official archives have invisibilized Black people. The field of Black feminisms is its own archive. We don't need official documents to prove how our folks have felt the university over time, rather we rely on our own "sacred 'libraries'" (Hundlle 2019, 304). We cannot think and speak and do from a place "of domination, subordination or 'subalternization,' but ... [from] slipperiness, elsewhereness" (Daives 1994, 36). Black feminism makes new life possible, sparks new questions, new research, new ideas and technological innovation, and also makes the possibility of physical departure for Black and Brown folks expressing a desire to leave less traumatic. It helps us make sense of our desire to leave in ways that go beyond "I don't feel well," or "I can't breathe." It is a place where we dance as a sign of refusal, "to refuse interpellation" (Halberstam 2013, 8–9), not because our feet are being shot at. According to Davies, "'elsewhere denotes movements," and "Black female subjectivity asserts agency as it crosses the borders, journeys, migrates and so re-claims as it re-asserts" (1994, 37). What I find powerful about the elsewhere and the whatever of Black feminisms is that, as Hong writes, it "does not concede the future to the present, but imagines it as something still in the balance, something that can be fought over, ... the work ... of imagining ... imagination is not frivolous or superficial activity, but rather a material and social practice toward 'revolutionary change'" (2008, 108). And yet, a Black feminist elsewhere also holds out that what we are doing is not future in the sense that Lee Edelman writes, about a politics of futurity which easily leads to an ethics of endless deferral (2004); such that we would remain in the grips of a deferred time which serves to consolidate the *status quo*. Focusing always on the better future, we would divert our attention from the here and now and thus be "rendered docile," in effect, "through our unwitting obedience to the future" (Kafer 2013, 29). Instead, a Black feminist elsewhere provides a place of freedom in the present: "the freedom of being, innately and complicatedly, a human being" (Quashie 2012, 36), who is not predetermined by socio-historical academic and ideological discourse, "but by desire, ambition, dreams, by one's affinity to the 'essences, the overtones, the tints, the shadows' of life as one takes it in ... being called from within" (Quashie 2012, 40) – from elsewhere.

Doing the shad(e)y work of Black feminisms presently imagines how Black feminist elsewheres are sites where we get better at our reading practices; sites where our Shad(e)y theoretics can be perfected so that we can let the institution know their shit ain't pretty, with their uses of empire to underwrite science, disciplinary knowledge, and epistemology in order to oppress. Our theoretics also guide us to "bring forth a poetics that envisions a decolonial future" (McKittrick 2013, 5), as through imaginative and creative acts we write "scientific and disciplinary knowledge anew, as necessarily a human project" (McKittrick 2015, 160). I touch on how we can create this project in the concluding chapter of this text by engaging more in-depth with Glissant's poetics of relation, which he writes, "remains forever conjectural and presupposes no ideological stability... against ... comfortable assurances ... open, multilingual in intention, directly in contact with everything possible" (1997, 32). I embrace conjecture to propose a different type of university space, one which allows us "to expand, to refuse, to open up space for our fullness rather than replicating the habits of enclosure" (Perry 2018, 233–234). In the next and concluding chapter, I propose the possibility of a creolized university.

References

Ahmed, Sara. *Strange Encounters: Embodies Others in Post-Coloniality.* London: Routledge, 2000.

Alexander, M. Jacqui. *Pedagogies of Crossing: Meditation on Feminism, Sexual Politics, Memory, and the Sacred.* Durham: Duke University Press, 2005.

Alexander, M. Jacqui. "Remembering This Bridge Called My Back, Remembering Ourselves." In *Feminist Solidarity at the Crossroads: Intersectional Women's Studies for Transracial Alliance*, edited by Kim Marie Vaz and Gary L. Lemons, 72–82. New York: Routledge, 2011.

Alexander, M. Jacqui and Chandra Talpade Mohanty. "Introduction: Genealogies, Legacies, Movements." *In Feminist Genealogies, Colonial Legacies, Democratic Futures*, edited by M. Jacqui Alexander and Chandra Talpade Mohanty, xiii–xlii. New York: Routledge, 2012.

Berry, Daina R. *The Price for Their Pound of Flesh: The Value of the Enslaved, from Womb to Grave, in the Building of a Nation.* Boston: Beacon Press, 2017.

Brah, Avtar. *Cartographies of Diaspora: Contesting Identities.* New York: Routledge, 1996.

Bravmann, Scott. *Queer Fictions of the Past: History, Culture, and Difference.* Cambridge: Cambridge University Press, 1997.

Byrd, Rudolph P., Johnetta Betsch Cole and Beverly Guy-Sheftall. *I Am Your Sister: Collected and Unpublished Writings of Audre Lorde.* New York: Oxford University Press, 2009.

Chatterjee, Piya and Sunaina Maira. "Introduction: The Imperial University." In *The Imperial University: Academic Repression and Scholarly Dissent.* Minneapolis: University of Minnesota Press, 2014.

Chen, Mel Y. *Animacies: Biopolitics, Racial Mattering, and Queer Affect.* Durham: Duke University Press, 2012.

Chinn, Sarah E. *Technology and the Logic of American Racism: A Cultural History of the Body as Evidence.* London: Continuum, 2000.

Christian, Barbara. "The Race for Theory." *The Nature and Context of Minority Discourse* Vol 6 (1987): 51–63.

Crawley, Ashton. "Introduction to the Academy and What Can Be Done?" *Journal of Critical Ethnic Studies Association* Vol. 4, no. 1 (2018): 4–19.

Davies, Carole Boyce. *Black Women, Writing and Identity: Migrations of the Subject.* New York: Routledge, 1994.

Davies, Carole Boyce. *Left of Karl Marx: The Political Life of Black Communist Claudia Jones.* Durham: Duke University Press, 2007.

Davis, Angela. *Freedom Is a Constant Struggle: Ferguson, Palestine, and the Foundations of a Movement.* Chicago: Haymarket Books, 2016.

Edelman, Lee. *No Future: Queer Theory and the Death Drive.* Durham: Duke University Press, 2004.

Fuentes, Marissa. *Dispossessed Lives: Enslaved Women, Violence, and the Archive.* Philadelphia: University of Pennsylvania Press, 2016.

Glissant, Édouard. "Creolization in the Making of the Americas." In *Race, Discourse, and the Origin of the Americas: A New World View*, edited by Vera Lawrence Hyatt and Rex Nettleford, 268–275. Washington: Smithsonian Institution of Press, 1995.

Glissant, Édouard. *Poetics of Relation*. Ann Arbor: University of Michigan Press, 1997.

Halberstam, Jack. "The Wild Beyond: With and for the Undercommons." In *The Undercommons: Fugitive Planning & Black Study*, 2–13. New York: Minor Compositions, 2013.

Hanchard, Michael. "Identity, Meaning and the African-American." *Social Text* no. 24 (1990): 31–42. Doi: 10.2307/827825.

Harney, Stefano and Fred Moten. *The Undercommons: Fugitive Planning & Black Study*. New York: Minor Compositions, 2013.

Hartman, Saidiya "Venus in Two Acts," *Small Axe* Vol. 26, no. 2 (2008): 1–14.

Hong, Grace Kyungwon. "'The Future of Our Worlds': Black Feminism and the Politics of Knowledge in the University under Globalization." *Meridians* Vol. 8, no. 2 (2008): 95–115.

hooks, bell. *Yearning: Race, Gender, and Cultural Politics*. Boston: South End Press, 1990.

hooks, bell. *Teaching to Transgress: Education as the Practice of Freedom*. London: Routledge, 1994.

hooks, bell. *Sisters of the Yam: Black Women and Self-Recovery*. Cambridge: South End Press, 2005.

Hughes, LeKeisha. "Robin D. G. Kelley and Fred Moten in Conversation Moderated by Afua Cooper and Rinaldo Walcott." *Journal of Critical Ethnic Studies Association* Vol. 4, no. 1 (2018): 154–172.

Hundle, Anneeth Kaur. "Decolonizing Diversity: The Transnational Politics of Minority Racial Difference." *Public Culture* Vol. 31, no. 2 (2019): 289–322. doi: 10.1215/08992363-7286837.

Johnson, Azeezat and Remi Joseph-Salisbury. "Are You Supposed to Be in Here?' Racial Microaggressions and Knowledge Production in Higher Education." In *Dismantling Race in Higher Education: Racism, Whiteness and Decolonizing the Academy*, edited by Jason Arday and Heidi Safia Mirza, 143–160. London: Palgrave Macmillan, 2018.

Kafer, Alison. *Feminist, Queer, Crip*. Indiana: Indiana University Press, 2013.

Livingston, Jennie. *Paris Is Burning*, 1991; Off-White Productions.

Lomax, Tamura. *Jezebel Unhinged: Loosing the Black Female Body in Religion and Culture*. Durham: Duke University Press, 2018.

Lorde, Audre. "The Uses of the Erotic: The Erotic as Power." In *Sister Outsider*. Berkeley, CA: Crossing Press, 1984.

Lorde, Audre, "Letter to Mary Daly." In *Sister Outsider: Essays and Speeches*. Berkeley, CA: Crossing Press, 1984.

Lorde, Audre "A Litany for Survival." *The Collected Poems of Audre Lorde*. New York: W.W. Norton & Company, Inc, 1997.

Lugones, María. "Playfulness, 'World'-Travelling, and Loving Perception." In *Making Face, Making Soul Haciendo Caras: Creative and Critical Perspectives of Feminists of Color*, edited by Gloria Anzaldúa, 390–402. San Francisco: Aunt Lute books, 1990.

Mbembe, Achille. *Decolonizing Knowledge and the Question of the Archive*. Africa is a Country, 2015. https://africaisacountry.atavist.com/decolonizing-knowledge-and-the-question-of-the-archive.

McKittrick, Katherine. *Demonic Grounds: Black Women and the Cartographies of Struggle*. Minnesota: University of Minnesota Press, 2006.

McKittrick, Katherine. "Plantation Futures." *Small Axe* Vol. 17, no. 3 (2013): 1–15. https://read.dukepress.edu/small-axe/article-abstract/17/3%20(42)/1/33296/ Plantation-Futures.

McKittrick, Katherine. "Yours in the Intellectual Struggle: Sylvia Wynter and the Realization of the Living." In *Sylvia Wynter: On Being Human as Praxis*, edited by Katherine McKittrick, 1–8. Durham: Duke University Press, 2015.

McKittrick, Katherine. "Axis, Bold as Love: On Sylvia Wynter, Jimi Hendrix, and the Promise of Science." In *Sylvia Wynter: On Being Human as Praxis*, edited by Katherine McKittrick, 142–163. Durham: Duke University Press, 2015.

Miller, sj and Nelson Rodriguez. "Introduction: The Critical Praxis of Queer Memoirs in Education." In *Educators Queering Academia: Critical Memoirs*, edited by sj Miller and Nelson M. Rodriguez, xv–xxiii. New York: Peter Lang, 2016.

Morgan, Joan. *When Chickenheads Come Home to Roost: A Hip-Hop Feminist Breaks It Down*. New York: Simon & Schuster, 2000.

Muñoz, José Esteban. *Disidentifications: Queers of Color and the Performance of Politics*. Minneapolis: University of Minnesota Press, 1999.

Myers, Joshua. "The Order of Disciplinarity, The Terms of Silence." *Journal of Critical Ethnic Studies Association* Vol. 4, no. 1 (2018): 23–43.

Nash, Jennifer C. *Black Feminism Reimagined: After Intersectionality*. North Carolina: Duke University Press, 2019.

Nero, Charles. *Towards a Black Gay Aesthetic: Signifying in Contemporary Black Gay Literature*. Michigan: ProQuest Information and Learning, 2005.

Perry, Imani. *Vexy Thing: On Gender and Liberation*. Durham: Duke University Press, 2018.

Quashie, Kevin. *The Sovereignty of Quiet: Beyond Resistance in Black Culture*. New Jersey: Rutgers University Press, 2012.

Sharpe, Christina. *In the Wake: On Blackness and Being*. Durham: Duke University Press, 2016.

Shaw, Andrea. *The Embodiment of Disobedience: Fat Black Women's Unruly Political Bodies*. New York: Lexington Books, 2006.

Simmonds, Felly Nkweto. "My Body, Myself: How Does a Black Woman Do Sociology?" In *Feminist Theory and the Body: A Reader*, edited by Janet Price and Margrit Shildrick, 50–63. London: Routledge, 1999.

Stallings, L.H. "A Eulogy for Black Women's Studies?" In *Feminist Solidarity at the Crossroads: Intersectional Women's Studies for Transracial Alliance*, edited by Kim Marie Vaz and Gary L. Lemons, 132–146. New York: Routledge, 2011.

Sultana, Farhana. "Decolonizing Development Education and the Pursuit of Social Justice." *Human Geography* Vol. 12, no. 3 (2019): 31–46. doi: 10.1177/194277861901200305.

Tallbear, Kim. "Making Love and Relations Beyond Settler Sex and Family." In *Making Kin Not Population*, edited by Adele Clarke and Donna Haraway, 145–163. Chicago: Prickly Paradigm Press, 2018.

Tinsley, Omise'eke Natasha. "Black Atlantic, Queer Atlantic: Queer Imaginings of the Middle Passage." *GLQ: Journal of Lesbian and Gay Studies* Vol. 14, no. 2–3 (2008): 191–215. doi: 10.1215/10642684-2007-030.

Vimalassery, Manu. "Fugitive Decolonization." *Theory & Event* Vol. 19, no. 4 (2016), muse.jhu.edu/article/633284.

Walcott, Rinaldo. "Multiculturalism, Cosmo-Politics, and the Caribbean Basin." In *Sylvia Wynter: On Being Human as Praxis*, edited by Katherine McKittrick, 183–202. Durham: Duke University Press, 2015.

Walcott, Rinaldo. "'Beyond the 'Nation Thing': Black Studies, Cultural Studies, and Diaspora Discourse (or the Post-Black Studies Moment)." In *Decolonizing the Academy: African Diaspora Studies*, edited by Carole Boyce Davies, Meredith Gadsby, Charles Peterson and Henrietta Williams, 107–124. Trenton: Africa World Press Inc, 2003.

Walker, Alice. *In Search of Our Mothers' Gardens: Womanist Prose*. San Diego, CA: Harcourt, 1983.

5 Conjecture not conclusion

Decoloniality, the poetics of science and curation ethics in "our" creolized elsewhere spaces

In the publishing world, the term "aliteracy" is used to refer to individuals who are able to read but refuse to do so. In her work, Kim Marie Vaz writes about "racial aliteracy" and "Racial alliterates." She writes "[i]n racial aliteracy African American culture is acknowledged and even 'celebrated,' but Whites appropriate aspects of African American culture to make it fit into a White worldview. This lack of sensitivity, empathy, and arrogance is a recognition and dismissal of 'the other.' Central to the ideology of racial aliteracy is this concept: It is not that the other does not exist but that the other exists for White purposes" (2012, 20). In Vaz's theorizing then, it is not that alliterates don't know how to read but that they purposefully engage in a misreading – a distortion of the lives and experiences of not just African Americans but Black people, "non-humans" and "non-persons" globally. In the previous chapter, I posited that if we are to achieve the potential of the elsewhere and the whatever, this intentional misreading must be countered by an even more intentional reading of the ugliness of this type of racial aliteracy. Throughout the text, I sought to not only read, but to also shade through telling, demonstrating that not only is the university ugly but also that it knows this and intentionally propagates this ugliness for its own futurity. Telling my own experience in the academy alongside the stories of those marginalized, this auto/biographical reading/telling serves as a "disruptive device that *reveals* my[/our] narrative as an *interpretive retelling*, vulnerable to challenge from other interpretations as the vagaries of self[/]-representations ... But the credibility of this narrative of political moments and events in dependent ... on the deeply invested self that speaks the events rel[ying] heavily upon the hope that its version will resonate with the meaning constructed by ... [our] various 'imagined communities'. My[/our] individual narration[s] ...[are] meaningful primarily as collective re-memory" (Brah 1996, 194). This telling, what I refer to as a Black feminist shad(e)y theoretics demonstrates the impossibility according to Édouard Glissant of "reduc[ing] anyone, no matter who, to a truth he would not have generated on his own. That is, within the opacity of his time and place" (1997, 194).

In the first two chapters of this book, I read how the scientific/academic inner workings of the academy have historically and contemporarily been

DOI: 10.4324/9781003019442-6

indifferent to the ways in which it has used technologies of oppressions to render Black people as non-human and women as non-persons, and the affective consequences that still resonate today. As universities continue to engage the liturgical performance of institutional "awareness" by pronouncements of "we acknowledge that we are on stolen land developed by enslaved people," I am brought back to the "we/us" that Nance, the parent mentioned in the introductory chapter to this book invokes, to again locate how these pronouncements carry within them the hidden meaning and the disavowal of historical and archival curation even as they are being spoken/performed. The ways in which the university archive has been curated to be understood as "just is," just showing us history as it unfolded, transparent and available for all to read the facts, holds within them the possibility for disruption via a good *read*. In this concluding chapter, I come back around to this read and ask several questions including; when we search for and find the elsewhere and the whatever what do we do next? In the context of the elsewhere and whatever, does science, the science that is built upon the rendering of Black people as flesh and women as nobodies, become irredeemable? And what will our curatorial practices, that is, the ways in which we preserve and care for our elsewhere places look like?

What we know from a queer and diasporic reading practice is that "[n]ot everything fits easily into a straightforward narrative. Narratives prune and require a structure that doesn't always account for all we need to attend to" (Perry 2018, 192-193). The historiography of the scientific narrative presented herein demonstrates this all too well. Acknowledging that this narrative is not straightforward means that when we engage in the traveling, and the reading, and the listening, and the looking in the last place they thought off, to uncover and reveal the elsewhere places, the non-transparent geographies of the ancestral, affective, communal, poetic, erotic, and the whatever else we might still yet be searching for and have not yet found, there is still work that remains to do. This work is the way in "which human beings create their relationships to their environment ... the idea of a larger "we," created by the truly poetic practice of developing a relationship with others that is not about negating the other to produce a human self but rather about the human poetry of creating and describing possible collective relationships to the environment" (Gumbs 2014, 240-241). If we are then to account for and make visible the possibilities of the elsewhere, concealed by colonization's calcifying effect, we must, according to Imani Perry, engage in a practice of curating our knowledges. The "we/our" that Perry writes about is important to what we curate and the ways in which we do curation. The we she writes about are the folks deliberately left out of the enlightenment knowledge project so that they/our ways of knowing show up in the academy as unscientific, unverifiable, incapable of being measured and therefore without merit. When we find the elsewhere places of our knowledge we must then be deliberate in how we engage such that we preserve them, not for neoliberal academic legitimation or acclaim but so as to create all types of signposts, like the underground railroad, as quilts intentionally dotting academic transparent landscapes visible to all but only knowable to us providing refuge and relation on a liberation journey. As Perry quoted in the previous chapter writes, we need to be deliberate in our curation practices. In order "to deliberately care for our souls and do so with discernment" (2018, 228–229).

She writes that our curation practices must be in the vein of bearing witness. The act of bearing witness brings to mind the reframe in Black religiosity "can I get a witness?" a reframe that calls forth the call and response the Black church uses. The call and response are melodic, a cultural expression the enslaved brought with them from Africa. The ways in which Black people have been able to curate the call and response such that it has become an important part of the Black institutions and oral traditions and is currently used to speak back against the machinations of being constructed as non-human and nonpersons in past and current abolition and liberation movements in the form of "what do we want? Justice! When do we want it? Now!" Show that the ways in which we curate, that is, organize and look after, that is care for our traditions as knowledge is important. It is important to both thinking about the pragmatism of Clare's two questions in Chapter 3, which are really questions about yearning – "How bad do you want this education? Are you willing to sacrifice things that make you comfortable?"– and the right to refusal such that our yearning for an education becomes shaped not entirely by the ideology of society but by cultivating the just imagination that Perry writes about or maybe an imagination for justice which instead of doing damage to self and others is about education "so that we can be transformed, by the passionate utterances, so that we can encounter other maps of human relations and the possibilities therein" (Perry 2018, 229). Paying attention to these call and response strands being uttered through a curatorial practice connects calls for social justice to desires for an education that will help us survive, will liberate – from poverty, from whatever we know and believe ails us and our communities. Connecting these strands through curation will help us to recognize how the current pragmatic posture to getting an education as one of surviving the longue durée of the present, can become transformed into a desire for education that is liberatory. If for us, liberation and "abolition means not simply the abolition of prisons but rather 'the abolition of a society that could have prisons, that could have slavery, that could have the wage,' then any university in the *after* of such abolition would be radically transformed, if not unrecognizable" (Stein 2018, 143-144). Curating, that is organizing, taking care of our cultural artifacts, means knowing that they exist, these whatevers of Black knowledge, and knowing where to look in the elsewhere, the last place they thought of, for these valuable artifacts of our knowledge toward liberation.

What does this curation look like? Quoting Perry again, "Curatorial practices are deliberate ways to be in relation to others in our midst – other people, but also other life forms, artifacts, and knowledge – with the understanding that we are being made as well as *being* in that process" (2018, 231). It does not mean curation in the sense of collecting and organizing for the white academic gaze or personal collections in a museum or gallery-like style, because even as Black cultural artifacts displayed in white museum spaces have been stolen from Black people, most of us don't see those spaces as ours but as an organization of our things for the white gaze. Riffing off Perry again our things "[w]e must recollect, not [as] a personal possession or egotistical drift, but as a call to ethics" (2018, 234). When I think about the academy and the ways in which students feel out of place in it, and the types of feelings this produces, I think about how our curatorial practices need to be ethical as a necessary counter to the deep and wide reach of the network of roots of the logics of the plantation still existing in our society today. We need to collect our things for ourselves such that despite and in spite of the existence of plantation logics "differential modes of survival emerge – creolization, the blues, maroonage, revolution, and more – revealing that the plantation, in both slave and post slave context, must be understood alongside complex negotiations of time, space, and terror" (McKittrick 2013, 3). This is a read!

McKittrick's use of the word creolization in the quoted passage above is instructive. Quoting Barbadian poet Edward Kamau Brathwaite, Rinaldo Walcott writes, "creolization is that which arises out of the brutal context and unequal power relations through which differing cultures come into contact and engagement with each other" (2015, 187). Walcott warns that we now inhabit a world that "accentuates an 'unknown creolization' that we cannot name but must struggle to recognize ... [and] require that we ask new questions" (2015, 188). The creolizing process is set in motion through conflict which arises from cultural contacts. It is a process that "opens on a radical new dimension of reality, not on a mechanical combination of components, characterized by value percentages. Therefore, creolization, which overlaps with linguistic production, does not produce a direct synthesis, but "resultantes," results: *something else, another way*" (Glissant 1995, 270). The creolization process is a poiesis, that is, a bringing into being what had not existed prior. There is no doubt that the creolization process was violent, but as Glissant writes it gave us jazz (1997, 73). Glissants reminds us that the cry/noise of the plantation was such as

> Night in the cabins gave birth to this other enormous silence from which music, inescapable, a murmur at first, finally burst out into this long shout – a music of reserved spirituality through which the body suddenly expresses itself. Monotonous chants, syncopated, broken by prohibitions, set free by the entire thrust of bodies, produced their language from one end of this world to the other, These musical expressions born

of silence: Negro spirituals and blues, persisting in towns and growing cities; jazz, *biguines,* and calypsos, bursting into barrios and shanty-towns; salsas and reggaes, assembled everything blunt and direct, pain-fully stifled, and patiently differed into varied speech. This was the cry of the Plantation, transfigured into the speech of the world (1997, 73).

I think about creolizing as a process of transformation birthed from sur-vival, as Walcott writes, it "is an altering of the human that concerns itself with surviving a process of "mutual mutations"" (2015, 188). People and cultures, according to Walcott, were "pitted against each other, or at least in tension with each other, while simultaneously living intimately with each other and sharing across those differences have produced "new" modes of being human in the region" (2015, 197) and I think about what this process can tell us or at least hold out for the university. That is, how can we think about the university as a space that can be transformed by a creolizing process, what would it mean to think about the university in terms of creolization rather than in terms of diversity and inclusion? How would Clare's statement "How bad do you want this education?" be turned on its head so that it is not one of pure practicality but also of possibility? How does thinking about the university in this way move us to a place where a Black feminist shad(e)y theoretics hold within it the possibilities of this elsewhere place which is a place beyond the university as we know it today, but that is achievable – an imaginary where we can feel for "a different future, for 'the future of our worlds' hangs in the balance" (Hong 2008, 108)?

Using the Caribbean as an example, I ask the readers to bear with me just a while longer as I engage in some "wild" conjecture. What if "the Caribbean basin as a space of cultural and identity experimentation has much to offer our thinking on how questions of identity, culture, ethics, and justice might inform ideas and practices of freedom"(Walcott 2015, 198) within the academy? What if we can imagine this? What does this look like? More importantly, what does this feel like? How do we get there - to a place where "our faculties ... [can] commingle reason and emotional in an ethical way ... [What types of] ... curation of study and thought" (Perry 2018, 243) would this require? I want to leave you with a few propositions.

First, the creolization process is an active and not passive one. As Maria Lugones writes, "We are *there creatively.* We are not passive" (1990, 400). As such, the creolization process is one that embraces chaos, in the way that both Lorde and Glissant write about chaos as "the opposite of what is ordinarily understood by 'chaotic' ... that ...opens onto a new phenomenon ... whose order is continually in flux and whose disorder one can imagine forever" (Glissant 1997, 133) and what I suppose this book is about. Lorde writes about the erotic as "a measure between the beginnings of our sense of self and the chaos of our strongest feelings" and Glissant about the "aesthetics of a Chaos, with every least detail as complex as the whole that cannot be reduced, simplified, or normalized" (1997, 32–33). Embracing then the possibility of creolization can be scary as it does not fall into the neat boxes of containment fashioned over time. By boxes I am referring to academic disciplinarily that require clear cut boundaries, an individual study that produces isolation. To realize the possibilities of a creolized university means complexity, chaos, disorder is necessary and must be embraced. As Glissant writes, "let us say that chaos is beautiful; not chaos born from hate and wars, but from extraordinary complexity of the exchange between cultures, which may yet forge future Americas that are at last and for the first time both deeply unified and truly diversified" (1995, 275). Thinking about chaos as generative rather than destructive reminds me of Harney and Moten's work on noise - "a cacophony and noise tells us that there is a wild beyond to the structures we inhabit and that inhabit us" (Halberstam 2013, 6) - music, and study.

In the book *The Undercommons: Fugitive Planning & Black Study* Fred Moten shares his thoughts about the study as

> what you do with other people. It's talking and walking around with other people, working, dancing, suffering, some irreducible convergence of all three, held under the name of speculative practice. The notion of a rehearsal – being in a kind of workshop, playing in a band, in a jam session, or old men sitting on a porch, or people working together in a factory – there are these various modes of activity. The point of calling it 'study' is to mark that the incessant and irreversible intellectuality of these activities is already present. These activities aren't ennobled by the fact that we now say, "oh, if you did these things in a certain way, you could be said to be have been studying." To do these things is to be involved in a kind of common intellectual practice. What's important is to recognize that that has been the case – because that recognition allows you to access a whole, varied, alternative history of thought (2013, 110).

Opening up the university in this way may seem chaotic, but at the same time, it helps us to recognize that studying is not and has never been "limited to the university. It's not held or contained within the university. Study has a relation to the university" (Harney and Moten 2013, 113). To focus on this opens up not only ways to be in relation creatively but also a series of disavowals of the we/us that have been constructed as belonging or not belonging to the university space and of the space as a purely rational one.

The beauty of chaos, or what Glissant refers to as *chaos-monde* that comes from a mixing is that what happens next is "unforeseeable and foretellable. We have not yet begun to calculate their consequences: the passive adoptions, irrevocable rejections, naïve beliefs, parallel lives, and the many forms of confrontation or consent, the many syntheses, surpassing, or returns, the many sudden outbursts of invention, born of impacts and breaking what has produced them, which compose the fluid, turbulent, stubborn, and possibly organized matter of our common destiny" (1997, 138). But being open to this chaos can produce all types of elsewhere spaces and whatevers. For example, I think about what Erica B. Edwards writes about embodied knowledges in "Sex after the Black Normal," when she writes "... *blues epistemology* ... the term [Clyde] Woods uses to describe 'an ethic of survival, subsistence, resistance, and affirmation' that sustains networks of labor and kinship throughout the Mississippi Delta region (*Development* 27). The embodied knowledge forms that comprise the blues tradition of explanation mark out geographies of surplus life, where those 'families, workers, and farmers classified [by social scientists] as tottering on the edge of 'extinction' outlive the rubrics that attempt to account for their lives and deaths ('Life' 62)" (2015, 156). How can this type of embodied knowledge and study from this elsewhere place that is not the university rub up against with chaos the spatiotemporal institution and what might be the creolized knowledge that comes from this that can produce our lives anew?

As I continue to engage in conjecture, of a creolization of the university, I also think about how it also requires a series of forgettings as well as remembrances. As Glissant writes about the creole language, it "has another, internal obligation: to renew itself in every instance of the basis of a series of forgettings. Forgetting, that is, integration, of what it starts from: the multiplicity of African languages on the one hand and European ones on the other, the nostalgia, finally for the Caribbean remains of these" (1997, 71). This forgetting, discarding, to reach a place of renewal must be done ethically. In his footnote on what forgetting means, Glissant also writes about the fragility entailed within it. He writes, "It is the problem of 'forgetting' that has made the various Creole dialects so fragile – in comparison to the languages composing them" (1997, 69). However, this is a risk that is necessary because the forgetting, casting off, maybe even rejection of some parts of the originating genres from which a creolized knowledge space emerges makes the creolized university not a shadow or copy of the neoliberal one such that these new chaotically created knowledges are not appendages to be included into old structures such that their inclusion is conditioned on hospitality, and in the furtherance of transparency and a racists sexist normality. Rather what emerges is a genre of knowing "with the force of a tradition that they built themselves, into the relation of cultures. ...But the truth is that their concern, its driving force and hidden design, is the derangement of the memory, which determines, along with imagination, our only way to tame time" (Glissant 1997, 71). According to Glissant *"creolization ... is an attempt at Being. ...We propose neither humanity's Being nor its models. We are prompted solely by the defining of our identities but by their relation to everything possible as well – the mutual mutations generated by this interplay of relations. Creolizations bring into Relation but not to universalize..."* (1997, 89).

Thinking about the future possibilities of the elsewhere spaces and the whatever in relation to the university in this way means that we do not have to jettison science. One of the questions that have been ruminating in my head as I write this book but that I have yet to address is, "is science (in the way it has been traditionally defined) irredeemable?" Has the violent history of science upon the bodies and Black and Brown folks and women made it such that we need to turn our backs on science altogether? In her essay "Axis, Bold as Love: On Sylvia Wynter, Jimi Hendrix, and the Promise of Science," McKittrick perhaps provides us a way of thinking about science that helps us to reconcile the ways in which its racist underbelly has kept us oppressed as we long for this scientific knowledge. McKittrick writes;

> the racial workings of science always already subjugate and/exclude marginalized communities, this bifurcating our *analytical approaches* to race, science, knowledge, and collaboration. It follows, then, that the creative works of black musicians, writers, and artists are distanced from, or simply unimaginable, in science studies and in the production of scientific knowledge. Yet in black studies, in addition to the research of Sylvia Wynter, the work of M. NourbeSe Philip, Aime Cesaire, Houston Baker Jr., Simone Browne, and Paul Gilroy, among others, explores such tangled scientific perspectives: black holes, DNA, infra-human categories, genomes, bloodlines, and poetic sciences and ana-lytical sites these thinkers utilize to work through racial politics and questions of emancipation. ... scientific racism cannot have the last word because this analytical frame refuses collaborative insights. While the natural sciences are certainly informed by monumental racial histories – and this is not to be dismissed – noticing conversations and connections between black creative texts and scientific knowledge will reveal important scholarly challenges: to breach analytical barriers and open up meaningful ways of imagining and honoring "a new contesta-tory image of the human" and therefore disclose otherwise unacknowl-edged political and intellectual narratives that *differently* imagine the scientific workings of emancipatory knowledge (2015, 149–150).

I think, for example, of Caribbean calypsonians like Shadow and the sci-ence of calypso music. Take, for example, how the Caribbean calypsonian engages methodically and with precision the double entendre, to express two sets of interpretations in the same line or verse, one meaning made plain and the other more implied and risqué. Also, the use of picong which as described by Kemlin Laurence refers to "'the exchange of teasing and even insulting repartee, generally in a light-hearted, bantering manner" (2017, 36). The poetic delivery of Caribbean calypsonian's lyrics requires a mastery and fusion of both the science behind music and study of cultural nuances – in the sense that Harney and Moten write about study – it is a powerful example of the way in which this creolization is both capable of provoking deep thought and feeling as it provides a commentary on our lived realities, but that is also ambiguous such that "poetics never culminates in some qualitative absolute" (Glissant 1997, 35).

"We are after a poetics, a practice, that unsettles, disorients, imagines otherwise possibility" (Crawley 2018, 11) and therefore scientific racism cannot have the final say in who we are and what we should desire, neither can the university that produces such science. Thinking about the university as a creolized space though recalls Stuart Hall words that, "The traces of ancient, stone-age ideas cannot be expunged. But neither is their influence and inflection permanent and immutable. The culture of an old empire is an imperialist culture; but that is not all it is, and these are not necessarily the only ideas in which to invent a future ... In the struggle for ideas, the battle for hearts and minds ... can only be displaced by better, more appropriate ones" (Hall 2017, 206).

Because creolization is a process of repetition, working through it will take time and effort, and may sometimes feel mundane and trivial. This repetition, though *"is an acknowledged form of consciousness both here and elsewhere"* (Glissant 1997, 45). To pull from and yet be present in the creation of something new, meaningful and just, *whatever* that ends up being is a deliberate practice "for one cannot improvise without practicing and arranging memory patterns developed through partly unconscious repetition and creative innovation" (McKittrick 2015, 159). This is I posit, the type of curation that Perry mentioned above calls us to do, to look deep within our knowledge archives and determine that our cultural artifacts are worthy of veneration, study, rumination, desire, and can provoke a different feeling or feeling differently in relation to not only ourselves but also with one another. Creating "a value system that imagines a version of being human that cannot be contained ... Put differently, the science of the word *feels and questions* the unsurvival of the condemned, thus dislodging black diasporic denigration from its "natural" place through *wording* the biological conditions of being human" (McKittrick 2015, 156). A creolized university is space where science and the poetics can exist not side by side but as a new creation altogether.

This is the type of decolonizing of the university I envision, "a state of being ... [in] persistent struggle by various peoples and institutions. Decolonizing is a process and it is relational, it requires solidarity networks to be built, as well as un/re-learning of accepted 'truths'" (Sultana 2019). I present this vision for the creolized university as a conjecture because in the words of Black feminist Barbara Christian we need "a tuned sensitivity to that which is alive and therefore cannot be known until it is known" (1987, 62). I present it from the elsewhere place of the crossing, a place which M. Jacqui Alexander describes as "that imaginary from which we dream the craft of a new compass. ... It is a place from which I[/we] navigate life" (2005, 8–9). This imaginary of the crossing has the potential for decolonial disruption as we engage "in the urgent task of configuring new ways of being and knowing and to plot the different metaphysics that are needed to move away from living alterity premised in difference to living intersubjectivity premised in relationality and solidarity" (Alexander 2005, 7–8). Black feminist Alexis Pauline Gumbs calls this "The poetics of survival, a queer relationality, is key to the practice of black feminist production" (2014, 254). To see "... blackness as a dialogue that can happen across space, without conforming to the structure of racism that places it in continually opposition to constructed whiteness" (Gumbs 2014, 254).

While what I am proposing here is not merely conjecture, it is a spatio-temporal reality I cannot know with certainty, but I can *read*, and because I *read*, I can feel, I can imagine. The poetics of creolization allows me to do that as "[th]e imaginative capacity inherent in poetry doesn't merely reflect the material world, nor is it an epiphenomenon of it, but rather is the 'skeleton architecture,' its 'foundation.' ... Lorde situated poetry as the *base*, not the superstructure. For Lorde, dismissing as 'luxury' the imaginative work of poetry has severe and bleak consequences: 'we give up the future of our worlds'" (Hong 2008, 108). The poetry that creolization can produce can build a new academy, one that is attuned and attentive to "Black diasporic wellness ... [where] all black flesh be interpreted as alive, as autonomous, as distinct, as spoken, as embodied, as feeling, as always already free, and inclusive, as in community, as preceding and exceeding contact/conquest, as self-determining, as power, as having survived, and as having a right to thrive – individually and collectively" (Lomax 2018, 206–207). I believe this, I desire this, I feel this, that "it is through poetry that we give name to those ideas which are, until the poem, nameless and formless, about to be birthed, but already felt" (Byrd, Cole and Guy-Sheftalll 2009, 185).

A Poem for Our Creolized Elsewhere Spaces

I am tired, tired of hearing "What don't kill you will make you stronger."
I don't want to die and I don't need no more strength
But I guess I have to die sometime, and isn't strength, well isn't strength a Black woman's curse?
Felt in our bones, it forecloses our dreams
But I dare to imagine something else, *whatever* you know?
I imaging it can be had, I feel it
I *aspire* it
Whatever moves my Black feminist theoretical artistry./?
Whatever making art from epistemological play/clay./?
Whatever fashioning chaos theory on a canvas layered poetically./?
Whatever ancestral!
Whatever feeling I am feeling!
Ah yes, Feeling that feeling of whatever!
Ah feelin ah feelin!

References

Alexander, M. Jacqui. *Pedagogies of Crossing: Meditation on Feminism, Sexual Politics, Memory, and the Sacred*. Durham: Duke University Press, 2005.

Brah, Avtar. *Cartographies of Diaspora: Contesting Identities*. New York: Routledge, 1996.

Byrd, Rudolph P., Johnetta Betsch Cole and Beverly Guy-Sheftall. *I Am Your Sister: Collected and Unpublished Writings of Audre Lorde*. New York: Oxford University Press, 2009.

Christian, Barbara. "The Race for Theory." *The Nature and Context of Minority Discourse* Vol. 6 (1987): 51–63.

Crawley, Ashton. "Introduction to the Academy and What Can Be Done?" *Journal of Critical Ethnic Studies Association* Vol. 4, no. 1 (2018): 4–19.

Edwards, Erica B. "Sex after the Black Normal." *Differences: A Journal of Feminist Cultural Studies* Vol. 26, no. 1 (2015): 141–167. doi: 10.1215/10407391-2880636.

Glissant, Édouard. "Creolization in the Making of the Americas." In *Race, Discourse, and the Origin of the Americas: A New World View*, edited by Vera Lawrence Hyatt and Rex Nettleford, 268–275. Washington: Smithsonian Institution of Press, 1995.

Glissant, Édouard. *Poetics of Relation*. Ann Arbor: University of Michigan Press, 1997.

Gumbs, Alexis Pauline. "Nobody Mean More: Black Feminist Pedagogy and Solidarity." In *The Imperial University: Academic Repression and Scholarly Dissent*, edited by Piya Chatterjee and Sunaina Maira, 237–259. Minneapolis: University of Minnesota Press, 2014.

Halberstam, Jack. "The Wild Beyond: With and for the Undercommons." In *The Undercommons: Fugitive Planning & Black Study*, 2–13. New York: Minor Compositions, 2013.

Hall, Stuart. "The Empire Strikes Back [written in 1982]." In *Political Writings: The Great Moving Right Show and Other Essays*, edited by Sally Davison, David Featherstone, Michael Rustin and Bill Schwarz, 200–206. Durham: Duke University Press, 2017.

Harney, Stefano and Fred Moten. *The Undercommons: Fugitive Planning & Black Study*. New York: Minor Compositions, 2013.

Hong, Grace Kyungwon. "The Future of Our Worlds": Black Feminism and the Politics of Knowledge in the University under Globalization." *Meridians* Vol. 8, no. 2 (2008): 95–115.

Laurence, Kemlin. "Trinidad English – The Origin of 'Mamaguy' and 'Picong.'" *Caribbean Quarterly* Vol. 17, no. 2 (2017): 36–39. doi: 10.1080/00086495.1971. 11829072.

Lomax, Tamura. *Jezebel Unhinged: Loosing the Black Female Body in Religion and Culture*. Durham: Duke University Press, 2018.

Lugones, María. "Playfulness, 'World' – Travelling, and Loving Perception." In *Making Face, Making Soul Haciendo Caras: Creative and Critical Perspectives of Feminists of Color*, edited by Gloria Anzaldúa, 390–402. San Francisco: Aunt Lute books, 1990.

McKittrick, Katherine. "Plantation Futures." *Small Axe* Vol. 17, no. 3 (2013): 1–15. https://read.dukepress.edu/small-axe/article-abstract/17/3%20(42)/1/33296/ Plantation-Futures.

McKittrick, Katherine. "Axis, Bold as Love: On Sylvia Wynter, Jimi Hendrix, and the Promise of Science." In *Sylvia Wynter: On Being Human as Praxis*, edited by Katherine McKittrick, 142–163. Durham: Duke University Press, 2015.

Perry, Imani. *Vexy Thing: On Gender and Liberation*. Durham: Duke University Press, 2018.

Stein, Sharon. "Higher Education and the I'm/possibility of Transformative Justice." *Journal of Critical Ethnic Studies Association* Vol. 4, no. 1 (2018): 130–153.

Sultana, Farhana. "Decolonizing Development Education and the Pursuit of Social Justice." *Human Geography* Vol. 12, no. 3 (2019): 31–46. doi: 10.1177/194277861901200305.

Vaz, Kim Marie. "Resegregating Women's Studies: 'Racial Aliteracy' – White Appropriation of Black Presences Revisited." In *Feminist Solidarity at the Crossroads: Intersectional Women's Studies for Transracial Alliance*, edited by Kim Marie Vaz and Gary L. Lemons, 19–31. New York: Routledge, 2012.

Walcott, Rinaldo. "Multiculturalism, Cosmo-Politics, and the Caribbean Basin." In *Sylvia Wynter: On Being Human as Praxis*, edited by Katherine McKittrick, 183–202. Durham: Duke University Press, 2015.

Index

Note: Page numbers followed by "n" refer to notes.

Printed in the United States
by Baker & Taylor Publisher Services

Printed in the United States
by Baker & Taylor Publisher Services